Applied Statistics with R

Applied Statistics with R

A Practical Guide for the Life Sciences

JUSTIN C. TOUCHON

Department of Biology, Vassar College, USA

OXFORD

UNIVERSITY PRESS

OXFORD
UNIVERSITY PRESS

Great Clarendon Street, Oxford, OX2 6DP,
United Kingdom

Oxford University Press is a department of the University of Oxford.
It furthers the University's objective of excellence in research, scholarship,
and education by publishing worldwide. Oxford is a registered trade mark of
Oxford University Press in the UK and in certain other countries

First Edition published in 2021

Impression: 1

Published in the United States of America by Oxford University Press
198 Madison Avenue, New York, NY 10016, United States of America

British Library Cataloguing in Publication Data

Data available

Library of Congress Control Number: 2021934831

ISBN 978–0–19–886997–9 (hbk.)
ISBN 978–0–19–886933–7 (pbk.)

DOI: 10.1093/oso/9780198869979.001.0001

Printed and bound by
CPI Group (UK) Ltd, Croydon, CR0 4YY

For Myra

Preface

The statistical analyses that life-scientists are being expected to perform are increasingly advanced and yet most graduate programs in the United States do not even offer a statistics course that teaches beyond Analysis of Variance (ANOVA) and linear regression. Undergraduate and graduate students are thus rarely provided with the opportunity to learn the types of analyses they need to know in order to publish and compete on the job market, much less simply analyze their data appropriately. Part of the reason for this is that the way statistics are traditionally taught can be frustratingly slow and tedious. When I was a graduate student, I remember excitedly enrolling in a statistics class with the hope of learning how to analyze the data I was collecting each summer in the field. Unfortunately, we spent the entire semester learning how to perform an analysis of variance and a linear regression, by hand. There has to be a better way!

This book is written with the belief that a comprehensive understanding of practical data analyses is not as daunting as it might seem. I have been teaching an annual statistics workshop at the Smithsonian Tropical Research Institute for more than 10 years and I know that my approach works. My teaching perspective is rooted in the idea that instead of spending time mired in statistical theory and learning data analysis by hand, the most important thing to understand is what kind of data you have. Once you know your data, you can then figure out how to analyze them effectively. Whether at the undergraduate, graduate, or post-graduate level, this book will provide the tools needed to properly analyze your data in an efficient, accessible, plainspoken, frank, and (hopefully) humorous manner, ensuring that readers come away with the knowledge of which analyses they should use and when they should use them.

This book uses the statistical language R, which is the choice of ecologists worldwide and is rapidly becoming the "go-to" stats program throughout the life sciences. The examples in the book are rooted in a single, real dataset (published in the journal *Ecology* in 2013) and use actual analyses that I have conducted in my professional career as an ecologist. The dataset is admittedly somewhat messy, and early chapters are designed so that students "clean" the raw data as a way of learning basic data manipulation skills and building good habits. Moreover, using a single relatively large dataset (~2500 observations) allows students to get a good understanding of what they are analyzing from chapter to chapter, instead of jumping from one small pre-cleaned dataset to another throughout the book. It also allows readers to see how they can view the same data through different lenses and allows an easy and natural progression from linear and generalized linear models to mixed effects versions of those same analyses, given the hierarchically nested design of the example experiment.

Goals for the book

It is my sincere hope that you find this book useful and instructive. I have tried my hardest to distill down everything I know and think about data analysis into these pages. You will undoubtedly find that some of what I suggest may differ from what you read elsewhere, either on the web or in other books. Just about everyone these days happens to be rather opinionated, and statisticians and R users are certainly no different. Wherever possible, I have tried to include the rationale behind my thinking.

Since you are reading this book, you evidently want to learn about data analysis. I applaud your initiative and to hope to reward you by teaching you how to do just that, efficiently and effectively. Here are the goals of this book.

- I hope to build your familiarity with R from the ground up via the chapters and assignments. Even if you have some experience with R, you will likely learn new ways to approach your data If you are relatively new to R, I hope the hands on experience of typing along with the instructions will help you overcome "fear of the R prompt."

- I want to empower you to not only follow instructions carefully and analyze the data presented in these chapters, but hopefully to be able to analyze your own data and to think critically about data when you see them presented in research and in the public realm. As you may already know, science literacy is seriously lacking in the public sphere and increasing the number of people who can think critically about data presented in the news or elsewhere is extremly important.

- Lastly, I hope you can become a part of the global R community. R is so big there is no single repository of information about it nor is there a single manual that contains all the possible instructions you might need to execute. Thus, in addition to books like this one, you will need to become familiar with using the web to find answers to questions. I will provide examples in the later chapters of how you might seek out information to help yourself when (not *if*, mind you, but *when*) you get stuck or encounter an error.

Basic layout of the book

The materials presented in these chapters are set up as follows. There are ten topics, each an explanatory chapter which will allow you to teach yourself the code. I cannot stress enough that you really do want to type things in and you need to think about what the code means and what it is doing if you want to learn this stuff. If you have an electronic copy of the book, avoid any temptation to cut and paste. If you are reading this, you are interested in learning R, right? Trust me, if you cut and paste code you *will not* learn as well as if you type it in by hand.

The ten topics are:

1 Introduction to R
2 Before You Begin (aka Thoughts on Proper Data Analysis)
3 Exploratory Data Analysis
4 The Basics of Plotting
5 Basic Statistical Analyses using R
6 More Linear Models!
7 Generalized Linear Models
8 Linear and Generalized Linear Mixed Effects Models
9 Data Wrangling and Advanced Plotting with the *tidyverse*
10 Writing Loops and Functions in R

Just a note about how each of the chapters will be formatted. Bits of code that you can/should type in are displayed in light grey boxes, and the output from that code is generally displayed directly below it. For example, check out the code below. What is shown in the grey box "*(2+2)*" is what you would type at the R prompt, and the bit of code below it is the output from executing that command.

```
2+2
```

```
## [1] 4
```

In general, if you type in exactly what is in the grey boxes you will get what is shown after it! Amazing, I know. Your mind is already blown, right?

The code that will be presented in this book is often written in a relatively "long" format in order to make it more readable. This might not exactly be how you type it to your computer though, which is perfectly fine.

At the end of each chapter is a short set of assignments to give you the opportunity to practice what you have just learned. You can find solutions to the assigments at the GitHub page for the book (https://github.com/jtouchon/Applied-Statistics-with-R) as well as other important informa-tion. Since R is an open source language it is likely that some of the code needed to run the examples in this book may change over time, and I will post code updates on that site.

A little background about R

R is a statistical programming package and a powerful graphics engine. R is considered to be a dialect of the S and S+ language that was created by AT&T Bell Labs. S is commercially available while R is open source and freely available through the Comprehensive R Archive Network: (https://cran.r-project.org). R has many advantages besides being freely available. For example, a user might program loops to conduct many repetitive statistical analyses or simulate thousands of data sets with known parameters. In addition, in the fields of Ecology and Evolutionary Biology at least, R is now by far the most commonly used statistical program (see Touchon and McCoy 2016 *Ecosphere*). There is substantial evidence that similar shifts are occurring in Psychology and Neuroscience as well.

A little about how R works

Because R creates objects from analyses that are stored in its memory, new users often are surprised by the fact that the results of their analyses are not immediately displayed on the screen. When you run something successfully, all you generally see is the prompt, which is denoted by the '>' sign.

There are several reasons for this. First, R does exactly what you tell it to do. Thus, if you tell it to run an ANOVA and store that output as an object, it does that, but you have to tell it a separate function to show you the object you created. Second, printing stuff on the screen takes time and computer power. By not showing everything that is going on, R is being very effcient. For example, if you wanted to do 100 regressions on different data sets, R can do this without opening 100 separate windows. One can store only the regression coefficients and display all of them in a single line for comparision. It is this flexibility that makes R a fantastic statistical program. Also, it's free. Did I mention that it is free yet?

This book provides an introduction to using R in data analyses with practical examples designed to be readily accessible to all life scientists. Although the example dataset I will use is ecological in nature, the parallels will hopefully be easy to see with other disciplines. A more explicit discussion of this is at the end of Chapter 2. R is also a very powerful graphing tool and I will get you started on your way to making publication quality figures.

This book is not a comprehensive overview of all available statistical approaches and methods or experimental design. No single book could do that. I will of course touch on many different topics, but there are over 16,000 packages available to use in R (as of July 2020), a number which is growing by the day, so such an overview is impossible.

Learning R is like learning any language. At times it will be diffcult and frustrating, but it is worth it and if you stick with it you will have breakthroughs that feel amazing (I call these "R-gasms"). Over time, you may grow to love working in R!

There is a quote I love from the musician, actor, author, poet, and all around amazing human Henry Rollins, which encapsulates a lot of how I think about doing statistical analyses and using R.

> Numbers are perfect, infallible and everlasting. You aren't. Numbers are always right in the end. You may see an incorrect figure, but that's not the fault of the number, the fault lies in the person doing the calculating.
> –Henry Rollins, *High Adventure in the Great Outdoors*

Why do I like that quote so much? It's because when you get an error in R, it is almost certainly your fault. R didn't mess up, you did. Sorry, but that's the honest truth. So check your code! :)

Why learn R?

You might be thinking to yourself "Why do I need to learn R?" or "Seriously, I have to type everything in by hand?!" or "Can't I do this easier in another program?" There are many answers to these questions.

- If you are an undergraduate thinking of going to graduate school, it is useful for you to learn R because you will almost certainly use R as a graduate student. Thus, you will have a leg up on everyone else! Get started now and be the best you can be.
- Yes, you have to type everything in, but that also helps you learn what you are doing. It is very easy to click some buttons and get an answer that you don't really understand. If you have to type in the code for the statistics you are doing, you will have a better understanding of what you are doing.
- Having some basic familiarity with "coding" is increasingly useful across a variety of disciplines. You don't need to be a pro, but being comfortable with a computer and with typing code to achieve a result is very useful.
- Because it is free and extremely powerful, R is the only statistics program you will ever really need to know. If you go on to graduate

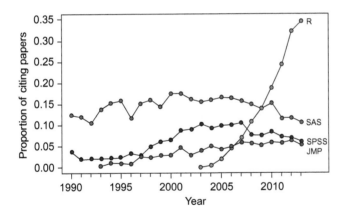

Figure 0.1 This figure, from Touchon and McCoy (2016), demonstrates the rise in usage of R as compared to SAS, SPSS, and JMP, in the field of ecology. R really is the go-to program, so it is in your best interest to learn it. Touchon, J.C. and McCoy, M.W. (2016). "The mismatch between current statistical practice and doctoral training in ecology." *Ecosphere.* 7(8):e01394. Reproduced under Creative Commons Attribution License (CC-BY)

school or into consulting or any field that deals with data, you will be able to use R. This book will teach you many of the basics you will need to know in R, but one of the best things about R is that it can be expanded to accomplish nearly any statistical (or, more generally, data analytic) needs you might have. The same cannot be said with other programs like JMP, SPSS, or SAS, which are very expensive and may not be available to you at another institution. Check out Figure 0.1 for evidence that R has become the program of choice (at least in Ecology, but the same is true in other fields as well).

Okay, shall we get started?

Acknowledgments

This book owes a tremendous debt to many people. First and foremost, thank you to Andy Jones and Stuart Dennis. The three of us took a germ of an idea—a desire to teach folks the practical tools they would need to analyze their data in R—and created the initial workshop at the Smithsonian Tropical Research Institute (STRI) that this material evolved from. Thank you to Owen McMillan, Adriana Bilgray, and Paola Gomez at STRI for their continued support of me and my desire to teach people how to use R. More generally, thank you to the amazing community of scientists at STRI for providing such an incredible environment to learn and conduct research. Many thanks to James Vonesh and Mike McCoy, two invaluable mentors, colleagues, and friends over the years. Your knowledge of R certainly eclipses mine, and I hope I've done justice to all that you have taught me. Thank you to my doctoral and post-doctoral advisor Karen Warkentin. Karen and James wrote the National Science Foundation grant that generated the data used throughout this book. Many thanks to Tim Thurman for opening my eyes to the world of *ggplot2* and *dplyr*. Thank you to the hundreds of interns, undergraduate, and graduate students, postdocs, and professional scientists that I have had the pleasure of teaching over the past decade or so. The lessons in this book have been continually refined and improved based on your feedback, so thank you for making me a better teacher. In particular, thank you to the students in my 2020

Applied Biostatistics class at Vassar College for the countless typos they found in early drafts of these chapters. Lastly, thank you to my wife Myra Hughey for her patience, support, and editorial advice over the years. You are the best partner in research and life I could ever hope for.

Contents

Preface   vii

Acknowledgments   xv

Chapter 1: Introduction to R   **1**

  1 Introduction to R   1
- 1.1 Overview   1
- 1.2 Getting started   2
- 1.3 Working from the script window   3
- 1.4 Creating well-documented and annotated code   4
- 1.5 Before we get started   9
- 1.6 Creating objects   9
- 1.7 Functions   13
- 1.8 What your data should look like before loading into R   21
- 1.9 Understanding various types of objects in R   23
- 1.10 A litany of useful functions   34
- 1.11 Assignment!   35

Chapter 2: Before You Begin (aka Thoughts on Proper Data Analysis)   **37**

  2 Before You Begin (aka Thoughts on Proper Data Analysis)   37
- 2.1 Overview   37
- 2.2 Basic principles of experimental design   37
- 2.3 Blocked experimental designs   39
- 2.4 You can (and should) plan your analyses before you have the data!   41
- 2.5 Best practices for data analysis   42
- 2.6 How to decide between competing analyses   44
- 2.7 Data are data are data   45

Chapter 3: Exploratory Data Analysis and Data Summarization   **49**

  3 Exploratory Data Analysis and Data Summarization   49
- 3.1 The Resource-by-Predation dataset   49

3.2 Reading in the data file   51
3.3 Data exploration and error checking   52
3.4 Summarizing and manipulating data   69
3.5 Assignment!   75

Chapter 4: Introduction to Plotting   **77**

4 Introduction to Plotting   77
4.1 Principles of effective figure making   77
4.2 Data exploration using *ggplot2*   80
4.3 Plotting your data   89
4.4 Assignment!   101

Chapter 5: Basic Statistical Analyses using R   **103**

5 Basic Statistical Analyses   103
5.1 Determining what type of analysis to do   103
5.2 Avoiding pseudoreplication   107
5.3 Testing for normality in your data   109
5.4 Non-parametric tests   117
5.5 Introducing linear models   124
5.6 One-way analysis of variance—ANOVA   125
5.7 Multiple comparisons   132
5.8 Assignment!   136

Chapter 6: More Linear Models in R!   **139**

6 More Linear Models!   139
6.1 Getting started   140
6.2 Multi-way Analysis of Variance—ANOVA   141
6.3 Linear regression   153
6.4 Analysis of covariance (ANCOVA)   163
6.5 The **predict()** function   168
6.6 Plotting with **ggplot()** instead of **qplot()**   174
6.7 Assignment!   179

Chapter 7: Generalized Linear Models (GLM)   **181**

7 Generalized Linear Models (GLM)   181
7.1 Understanding non-normal data   181
7.2 GLMs   183
7.3 Understanding and interpreting the GLM   188
7.4 Calculating statistical significance with GLMs 195
7.5 Coding the data as a binomial GLM   198
7.6 Mixing GLMs and ANCOVAs together   200
7.7 Using the **predict()** function with a GLM   204

7.8   Making a much easier GLM/ANCOVA plot
        using ***ggplot2***   206
7.9   Assignment!   208

Chapter 8: Mixed Effects Models   **209**

8   Mixed Effects Models   209
8.1   Understanding mixed effects models   209
8.2   Assignment!   233

Chapter 9: Advanced Data Wrangling and Plotting   **235**

9   Advanced Data Wrangling and Plotting   235
9.1   The "tidyverse"   235
9.2   Basic data wrangling   238
9.3   Advanced data wrangling: Spreading and gathering your data   245
9.4   Even more advanced data wrangling!
        Using the **do**() function   249
9.5   Making better figures with ***ggplot2***   256
9.6   Basics of ***ggplot2***   257
9.7   Customizing your figure   267
9.8   Combining data wrangling with plotting
        with ***ggplot2***   274
9.9   Assignment!   282

Chapter 10: Writing Loops and Functions in R   **285**

10   Writing Loops and Functions in R   285
10.1 for loops   286
10.2 Understanding functions   288
10.3 Writing functions   289
10.4 How a function works   289
10.5 Writing more complex functions: An example using simulations 292
10.6 Assignment!   306

Chapter 11: Final Thoughts   **307**

11   Final Thoughts   307
11.1 Understanding your data is the most important precursor to
        analyzing it   307
11.2 Knowing how to get help is essential   308
11.3 Your data analysis should be clear from the outset and you should
        avoid questionable techniques   308
11.4 Presenting your data in well-constructed figures is key   309

Index   311

Introduction to R

1 Introduction to R

1.1 OVERVIEW

The purpose of this first chapter is to introduce you to the basic workings of R and get you up to speed. Some of this material might be familiar to you if you've used R before, but the goal is to get anyone reading the book up to a basic level of familiarity. You will learn many of the basic and very important functions of R, such as:

- Creating objects
- Writing articulate R code
- Using functions
- Generating artificial data
- Entering data in a format that can be read and analyzed by R

This chapter does not intend to be an exhaustive introduction to all the basic workings of R. In other words, we'll move pretty quickly here. If you would like a greater introduction, I highly recommend checking out the

Applied Statistics with R: A Practical Guide for the Life Sciences. Justin C. Touchon, Oxford University Press (2021). © Justin C. Touchon. DOI: 10.1093/oso/9780198869979.003.0001

excellent book *Getting Started With R: An Introduction for Biologists* by Andrew Beckerman, Dylan Childs, and Owen Petchey.

1.2 GETTING STARTED

1.2.1 Obtaining R

If you are brand new to this, you will have to download R in order to do anything. Just navigate your web browser of choice to http://cran.r-project.org to download the appropriate version of R for your operating system. There is another program you may have heard of and may want to use called RStudio, which can be found at http://www.rstudio.com.

Box 1.1 - RStudio

Please remember this: RStudio is a program that uses R. It helps keep things organized and has some nice autocomplete functions, but R is the actual program that does *everything* that we will cover in this book. RStudio has plenty of great features, don't get me wrong. It's really great for writing in RMarkdown and LaTeX, if you choose to do that. But, like I said, RStudio is a program that uses R. R does all the heavy lifting. R is the statistics program. Personally, I use regular plain old R and not RStudio. To each there own ….

1.2.2 Installing and loading packages

R is designed to be a small program (currently just about 80 mb) which makes it easy to download and install anywhere in the world. The base version of R contains a great number of functions for organizing and analyzing data, but the real strength comes in what are called ***packages***. Packages are freely downloadable additions to R that provide new functions and datasets for particular analyses. For example, the base version of R can conduct linear models and generalized linear models (Chapters 5 – 7) but cannot conduct mixed effects models (Chapter 8). To do mixed effects models, you need to download a specific package (of which there are several).

The only important thing to remember about packages is that adding them to R is a two-step process. First, you have to **install** a package, which (perhaps counterintuitively) just downloads the package to your computer.

Secondly, you have to **load** the package, which is when you have actively placed it in the current memory for use. You will generally obtain packages from the Comprehensive R Archive Network (https://cran.r-project.org/) (CRAN) directly through R.

Box 1.2 - Install some packages

Assuming you have installed R on your computer, you should run the following code to install the various packages you will need to have in order to execute the commands presented throughout this book. If you are using RStudio you can click on the packages tab and search for these one at time by using the little search window. Make sure to click the button to "Install Dependencies." We won't do anything with these right now, but they will be necessary later in the book.

```
install.packages(c("lme4","multcomp","car","ggplot2","gplots",
            "MASS","tidyr","dplyr","broom","gridExtra",
            "cowplot","emmeans","glmmTMB","lattice"),
            dependencies = T, repos =
            "http://cran.us.r-project.org")
```

1.3 WORKING FROM THE SCRIPT WINDOW

The biggest mistake that most new R users make is to just type commands into the command prompt. The problem with this is that once you hit enter the command is gone. If you hit the up-arrow, R will scroll through the previously executed commands, but aside from this what you typed is gone and *it cannot be edited*! It is of course reasonable to run lines from the command line from time to time, but it is much better to work from a script window.

The script window allows you to easily save and edit your code, and to execute one or multiple lines of code at once. To open a blank script window, go to the File menu and click on New Document, or just hit command-N (Mac) or control-N (PC) on your keyboard.

In the script window you can type in your commands and then execute them by hitting command-enter (Mac) or control-R (PC). This means you type code into the script window and then the program sends the line of code to the command prompt for you. Do not cut and paste code from the script window to the command prompt; that is a waste of time. You can also highlight multiple lines of code and execute them all at once. To save your code simply go to the File menu and save as you would any other file (or just hit command-S or control-S on your keyboard).

A script allows you to edit, run, and tweak your code, save it, return to it later, send it collaborators or mentors, and so on. Anything you think will want to run more than once, or that you might want to edit, should be typed into a script window (which is pretty much everything).

1.4 CREATING WELL-DOCUMENTED AND ANNOTATED CODE

One of the most important things you can do is write orderly, well-annotated code that not only functions well but explains what is happening and why it is happening and does so in easy to read and understand language. This idea was first introduced by computer scientist Donald Knuth and is known as "literate programming." Literate programming is the process of interspersing your computer code, in this case R code, with plain-language descriptions of what the code is doing. This allows a reader to have a fully formed idea of what is going on. In R, you do this with annotation, which is simply the process of leaving notes within the code that are not actually code themselves. It's like you are Hansel and Gretel getting dragged into the woods: you want to leave plenty of clues for your future self (or others) to be able to discern the trail you took.

Box 1.3 - Write good code, for yourself and others

For any bit of R code you write, you should consider that you are writing for three audiences:

1) Your future self
2) Your collaborators
3) Everyone else that might look at your code one day

---

1. Writing for yourself

Seldom (never?) will you have the opportunity to sit down with a dataset and analyze it start to finish in a single sitting. It is rare that you even will have the opportunity to work on it on consecutive days where what you did yesterday is still fresh in your mind today. What is more realistic is that you work on something for some period of time (hours, days, maybe even weeks if you are really lucky!) then have to put it down for some time because you are distracted by other tasks (teaching, other research demands, manuscript revisions, parenting, a pandemic, etc.). By the time you come back to your code even a week later, you will likely have to invest some substantial time getting back to where you were. Writing good, clear code will reduce that restart time considerably.

2. Writing for collaborators

If you are, or are planning to be, a professional scientist, you are unlikely to work exclusively by yourself. There will be times when you collaborate with others. Maybe it's your graduate advisor, maybe a colleague at another institution. Whatever the scenario, it means you might be responsible for analyzing or organizing some set of data, then sharing it with others. If that's the case, you want to make sure when you send your code it is clear what you did and why you did it. Imagine the embarrassment of your collaborator sending you question after question trying to figure out what your code means!

3. Writing for folks in the future who might want to see your code

Increasingly, it is necessary to post both the data that go into a scientific article and also the code that was used to analyze it. This is a tremendous step towards increasing transparency in science and is to be applauded for sure. But it also means that some stranger might look at your code a month or a year or more down the road, even after you thought you were long done with it all. Thus, just like writing for your future self or your collaborators, you want to make sure that your code is clean and organized and well-annotated.

Box 1.4 - How do you create good code?

There are a few basic rules of thumb that you should do your best to adhere to. These will help ensure you stay organized and help prevent embarrassment when you inevitably have to show your script file to someone else! It will also just make your life in R much, much easier.

1. Always start with your raw data
2. Annotate, annotate, annotate
3. Organize your script file into logical sections
4. Give your objects meaningful and unique names

---

1. Always start with your raw data
Import your raw file from Excel or whatever program you use for data entry but do whatever data processing is necessary (removing outliers, calculating new variables from old ones, etc.) within an R script. This way, you have a record that others can follow that leads them from the raw data to the final product. This also increases transparency in science: if you do a bunch of data filtration before you import your data, you are omitting potentially important information from the scientific record.

2. Annotate, annotate, annotate
Leave lots of notes to yourself or others about what certain bits of code are doing and the rationale behind them. You annotate your code by using a number sign, also called a pound sign (or hashtag, as the kids might say). Whatever you call it, it's this thing: #. Anything that is written after a # on a given line won't be executed. For example, consider the following two bits of code (feel free to type this into your own R console and see for yourself what happens).

```
Howdy
```

```
## Error in eval(expr, envir, enclos): object 'Howdy' not found
```

Note that we got an error because R doesn't know what this word is supposed to mean. R assumes that *"Howdy"* must be an object in the memory, but it is not there, so it gives us an error. Now see what happens if we put a "#" in front of the word.

```
#Howdy
```

What happened? Nothing, absolutely nothing! The line was not executed because it came after a #. This simple coding trick allows you to leave yourself descriptive notes in your code that won't interrupt the executable code. Throughout this book, you will see notes embedded in the chunks of code that help describe what is going on.

3. Organize your script file into logical sections

I like to use long strings of #s to create large visible breaks in my script files that I can see when I'm quickly scrolling through it. At the top of the script file, I will put a header, each line starting of course with #s, that describes the basics of what the file is, where the data are from, the date I started working on it, that sort of thing. Then, I create a section where I load all the packages I will need for that script file. You can use a # and title it something like *"#Packages to load."* Then, create whatever other sections you might need for your purposes. These might include sections for analyzing different aspects of the data, or running different types of analyses, or to split apart analyses vs figure making.

Box 1.5 - An example of a script header

I like to give my code a nice big clear header at the top which describes what that particular file is for. I also like to put a block of code that loads whatever packages I will need for that particular set of analyses. Here is an example of what that might look like.

```
##########################################
###Code for analyzing RxP experiment data
###Date created: Nov. 15, 2012

###Load the packages
library(dplyr)
library(ggplot2)
library(lme4)
library(emmeans)
library(MASS)
```

4. Give your objects meaningful and unique names

As my good friend Stu Dennis used to say, "You wouldn't call your dog 'dog,' so don't call your data 'data.'" I couldn't agree more. Make sure when you are importing your dataset, creating objects, running and saving models, etc., that you assign useful and meaningful names. The flip side of this is that you generally want those names to be short, since typing in a lengthy object name over and over gets a little laborious. The same logic applies to the names of variables in your data frame. For example, in the dataset we will work with in this book, there is a variable that represents the snout-vent length of red-eyed treefrog froglets at the start of metamorphosis, which is referred to as simply "SVL.initial" in the dataset. It's short enough that you don't mind typing it, but unique enough to let you know what it refers to.

You can customize all of this for your own purposes, and it is a good idea to try to be consistent throughout your coding. For example, in the previous paragraph I used a period (.) to take the place of a space, since R generally does not like spaces. Alternatively, I could have used an underscore (_) or a hyphen (-), thus creating the hypothetical variables "SVL_initial" or "SVL-initial." Any of these are acceptable, the key is just to be consistent with how you create spaces. Similarly, decide if you want to capitalize variable names or not and stick to it. Remember, R will distinguish between uppercase and lowercase characters (e.g., *SVL.initial* is different from *svl.initial*), so being consistent will help you prevent frustrating mistakes.

The basic point is to just try to be clear and organized and leave instructions for yourself and others. There are many excellent resources out there that provide a much more exhaustive take on this subject than I have space to provide here, and I certainly encourage you to seek them out. One that I am quite fond of is the British Ecological Society's *A Guide to Reproducible Code in Ecology and Evolution*. A quick internet search will help you find it.

1.5 BEFORE WE GET STARTED

Before we begin, there are just a couple of things that are really useful to know about R and how it works. When you see the greater than sign (">") in the console, that means R is ready for you to type something in. If instead you ever see a plus sign ("+"), that means R is waiting for you to finish some command. R generally knows when a command has been finished or not. If you type into the console "2+", R is going to wonder what you are adding to that first 2. You can then type something at the "+" prompt to finish the command you are trying to execute. You can also hit the escape key to cancel whatever command R is waiting for you to finish. In the examples of the code in this book, you won't see the little greater than sign that you see in your console window.

```
#An example of an unfinished command
> 2+
+
```

This works to our advantage when we are writing out more complex things like statistical models. If we start a function, like **lm()** to run a linear model (Chapters 5 and 6), R will keep the command open until we close the parentheses and finish out the command. If that doesn't make sense just yet don't worry, it will.

1.6 CREATING OBJECTS

Let's start with creating simple **objects** in R. Objects allow the user to create very simple symbolic representations of simple or complex datasets or other information which allows the user to then create and run elaborate analyses from seemingly simple code with the object stored in the computer's memory. However, creating and manipulating objects can be hard to get used to for those that are accustomed to drag and drop menus or using spreadsheets to manually manipulate and analyze data by selecting columns or individual cells. In the end, however, creating and manipulating objects is far more efficient than manually manipulating data, as I hope you will see by the end of the chapter.

Objects are created with the **assign** operator, which is an arrow made from a less than sign and a minus sign and looks like "<-". For example, if we want to create an object called **n** and assign the value of 10 to it, we would type the following:

```
n <- 10
```

In R lingo, you can say this as "n gets 10" or "10 is assigned to n." It can go the other way as well, although this is less common.

```
10 -> n
```

In both of these examples, you created an object called **n** and assigned it the value of 10. You may have noticed that nothing happened after you hit return. In general, that is good. You told R to do something and it did it. You did not tell R to display what **n** was. To do that, you should just type **n** at the command prompt.

```
n
```

```
## [1] 10
```

Remember earlier when we typed the word "*Howdy*" at the prompt and got an error? Well, if we assign something to be called "*Howdy*" then it will be stored in the memory and we can access it.

```
#Assign a value to be stored as Howdy
Howdy <- 23
#See what happens when you type Howdy into the prompt
Howdy
```

```
## [1] 23
```

You can also use the = sign to do this (e.g., *n = 10*). Some folks prefer the equal sign because it is one less key stroke, others prefer the little arrow. I recommend the arrow because it separates assigning objects from other uses of the equal sign, and that is what you will see in these chapters. Once again, to each their own.

Names of objects *MUST* start with a letter and can include letters, numbers, dots (.), and underscores (_). R is case sensitive, so "x" and "X" are different objects. Remember, spelling always counts and spelling mistakes are among the most frequent "bugs" in R code. Whenever you get an error, the first thing you should always look for is a typo in your code.

> **Box 1.6 - Always remember the following!**
>
> R does exactly what you tell it to do and only what you tell it to do.
> There is no spell check and no autocorrect.
> Spelling counts.
> R doesn't make mistakes, we make mistakes.

One thing to be careful of is that it is very easy to write over and replace existing objects. For example, earlier we made an object called **n** and assigned it the value 10. We can easily write over the first **n** with a new version assigned a different value.

```
n <- 15
n
```

```
## [1] 15
```

Maybe it is obvious, but this is a really important thing to realize. Note that when we overwrite **n** with a new value R doesn't give us any warning or anything to say "Hey, you already have something called n in the memory! Are you sure you want to do this?" You could easily overwrite your whole dataset with a few keystrokes and R won't blink an eye (not that R has an eye to blink, because it is a computer program).

We can also use R as a calculator.

```
2+2
```

```
## [1] 4
```

```
2*2+1
```

```
## [1] 5
```

```
2*(2+1)
```

```
## [1] 6
```

```
2^4/2+1
```

```
## [1] 9
```

```
2^(4/2)+1
```

```
## [1] 5
```

```
2^(4/(2+1))
```

```
## [1] 2.519842
```

Objects stored in the memory can of course also be used in math calculations. Since we have assigned numerical values to both **n** and **Howdy**, we can add them together.

```
n + Howdy
```

```
## [1] 38
```

1.7 FUNCTIONS

Most of the work done by R is going to be done by functions. In Chapter 10 you will learn how to write your own functions, but for now let's just discuss what they are. Functions are pre-written sets of code that we use to do something to a numerical value or set of values organized into an object. Functions can be very simple, like calculating the average of a set of numbers, or very complicated. Functions have two basic parts: a name and a set of parentheses where you specify the objects or values you want to the function to work on. Everything you type between the parentheses are called **arguments**.

For example, the **rnorm()** function calculates a random number selected from a normal distribution. In its simplest form, we can just pass it a single number and see what happens.

```
rnorm(1)
```

```
## [1] 0.565837
```

That produced a single value, which might be positive or negative and is not too far from zero. What happens if you change the 1 to a different number? What happens if you execute the function a bunch of times?

If you run that function over and over, you will notice a pattern and you might start to wonder why the random values being generated are so close to zero. There are a number of hidden default values for the **rnorm()** function. How do we know what they are? This is a good time to introduce the **help** function, which is just a simple **?**. If you are using RStudio you can click on the help tab (lower right screen) and type in your search term there, but the quickest way is to just type a **?** at the prompt followed by the thing you are looking for, like so:

```
?rnorm
```

This brings up the help file for the function, which describes a variety of functions related to the normal distribution. In this case, we are only interested in **rnorm()**. Note that interpreting a help file is a skill that you develop over time. The first time you look at one, it probably looks like gibberish. Don't worry, you'll get better at it.

Every function takes what are called **arguments**. Some arguments are required for you to enter in order to make the function work, others have default values that the function will use if you do not specify anything different. The description of the **rnorm()** function shows three arguments (Figure 1.1). The first value, "n," is the number of observations, or how many values you want the function to return to you. It has no predefined value and so the user (i.e., you) must enter a number to tell the function how many values to return. The "mean" and "sd" are the mean and standard deviation of the normal distribution that R will use when generating the requested random values. By looking at the help file, we can see that the pre-assigned values are a mean of 0 and a standard deviation of 1. However, we can change those values and draw any number of observations from any normal distribution.

For example, if we want to generate 100 random numbers drawn from a distribution with a mean of 5 and a standard deviation of 2, we can type the following code. Notice that I assigned the output of the function to be an object called *r1*.

```
r1<-rnorm(100,5,2)
r1
```

```
##    [1] 3.7625033 7.6065803 6.3051067 7.6477668 2.5110378
##    [6] 5.0259595 9.1432082 2.3651186 6.4309443 8.9676098
##   [11] 6.8264936 4.1375613 5.5034346 5.9325743 7.7151804
##   [16] 8.0669680 5.7715081 3.6696001 4.3587685 4.6954849
##   [21] 3.4907821 2.4837913 8.0522636 5.7626055 4.4145358
##   [26] 6.0727592 4.7932313 6.0531992 7.3823991 0.3500242
##   [31] 3.7008020 1.5271106 6.2922795 4.2418119 3.1913386
##   [36] 1.3537128 2.3712010 4.3589197 5.1556329 6.4521774
##   [41] 8.7622133 6.5129916 6.3680607 5.2462455 4.0613280
```

```
##   [46]  5.5311980  1.5841967  3.3453341  0.3775412  6.4556648
##   [51]  4.9400145  3.7575483  5.1898603  7.1110720  5.1301011
##   [56]  6.3759847  9.2248123  3.9554938  8.8057606  1.9586191
##   [61]  7.6720783  7.0384348  8.2352965  6.6933817  5.5923564
##   [66]  8.2956376  3.5567489  4.6238918  7.4281710  7.4639680
##   [71]  5.5056979  4.8943681  5.9444343  4.5263508  8.4303802
##   [76]  3.7098617  4.5443020  2.2479513  6.1758213  3.5201645
##   [81]  8.5453996  3.4135835  7.4039578  2.9011293  4.3179926
##   [86]  3.4551065  4.5932194  4.4821694  8.2074365  4.4339075
##   [91]  4.4112385  4.7239169  4.0373923  1.7402423  5.8637959
##   [96]  5.6873017  9.2210731  6.5497683  6.5620437  4.6091549
```

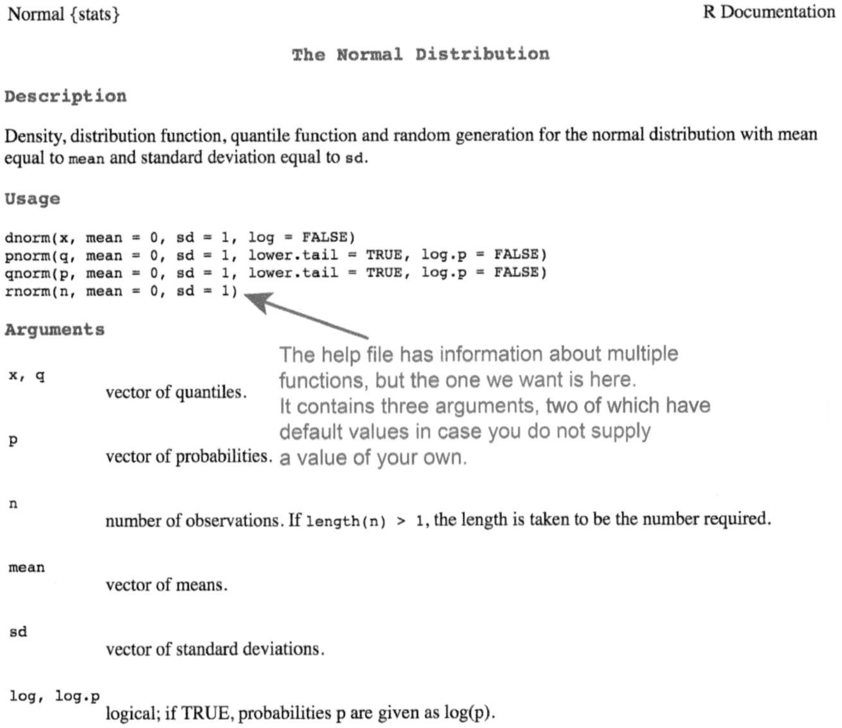

Normal {stats}   R Documentation

The Normal Distribution

Description

Density, distribution function, quantile function and random generation for the normal distribution with mean equal to mean and standard deviation equal to sd.

Usage

```
dnorm(x, mean = 0, sd = 1, log = FALSE)
pnorm(q, mean = 0, sd = 1, lower.tail = TRUE, log.p = FALSE)
qnorm(p, mean = 0, sd = 1, lower.tail = TRUE, log.p = FALSE)
rnorm(n, mean = 0, sd = 1)
```

Arguments

x, q
  vector of quantiles.

The help file has information about multiple functions, but the one we want is here. It contains three arguments, two of which have default values in case you do not supply a value of your own.

p
  vector of probabilities.

n
  number of observations. If length(n) > 1, the length is taken to be the number required.

mean
  vector of means.

sd
  vector of standard deviations.

log, log.p
  logical; if TRUE, probabilities p are given as log(p).

Figure 1.1   This is the help file for the function **rnorm()**. Notice that there are actually four different functions listed at the top of the file, all of which are related to the normal distribution. The help file shows us that there are three arguments that can be passed to **rnorm()** but two of those arguments have default values, which will be used unless you specify something else.

Note that your values for **r1** will be different than what is shown here, since the values are drawn randomly. This provides the opportunity to mention a couple of important things. First, we assigned the output of **rnorm()** to the object **r1**. There is nothing special about the name r1, I just picked it out of thin air. You could have called it "Carrie" or "Skywalker" if you wanted. In general, it's a good idea to keep your object names short since you have to type them in. More importantly, instead of having all 100 values spit out in the console, where they would be generated once and then are gone, we have stored them as a **vector** (we will discuss vectors more in just a bit). We can determine the number of objects in a vector using the function **length()**.

```
length(r1)
```

```
## [1] 100
```

Second, you will notice that at the start of each line there is a number in brackets. Each item, or **element**, in the vector has a place, or a number (1, 2, 3, etc.). The vector wraps around in the console and the numbers you see on the left side of the output is where in the vector the line has wrapped. We can call any one or multiple elements of a vector by using the index function, which is denoted by "[]" (square brackets). Square brackets are actually a type of function and indexing is a fundamental and key part of understanding how to use R to access and manipulate objects. For example, assume I want the first element of **r1**, or the tenth object, or whatever you can imagine.

```
r1[1]
```

```
## [1] 6.870637
```

```
r1[10]
```

```
## [1] 4.267947
```

We can also do this for multiple elements. For example, if we write two numbers with a colon between them, R will make a range of whole numbers from the first to the last. Thus, if we wanted to get objects 5 through 10 in vector **r1**, we can type the following.

```
r1[5:10]
```

```
## [1] 8.624091 3.429006 2.169269 6.095943 5.412810 4.267947
```

The "*5:10*" we typed previously is actually a shortcut to create a vector of numbers between 5 and 10. Go on, try it out. Type any two numbers at the console separated with a colon. What happens if you make the first number larger than the second number? Similarly, we can specify exactly which objects we want by putting a vector of numbers inside the square brackets. We can make a vector of numbers, letters, etc., using the concatenate function, which is just **c()**. Thus, if we wanted to get objects 10, 23, and 65 (for example), we would type this.

```
r1[c(10,23,65)]
```

```
## [1]  4.267947 5.448676 4.512647
```

We will come back to indexing and vectors later but keep them in mind. A couple of other useful functions. To see what you have stored in the memory, use the **ls()** command (short for "list").

```
ls()
```

```
## [1] "Howdy" "n"       "r1"
```

To remove an object, use **rm()** (short for "remove").

```
rm(n)
```

We can do some quick visualization of the data in **r1** using the **hist()** function (see Figure 1.2).

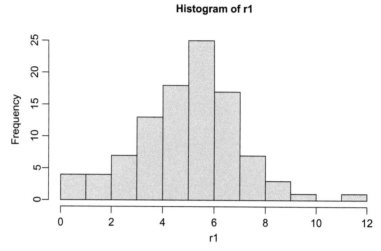

Figure 1.2 A histogram of 100 randomly generated values from a normal distribution with mean of 5 and a standard deviation of 2. Note that your histogram will look a little different because you have 100 different randomly generated values than I had when I made this figure.

```
hist(r1)
```

> **Box 1.7 - Functions within functions within functions**
>
> You can also embed functions within functions. Here are some things to try:
>
> 1. Make a histogram of 1000 random variables with a mean of 10 and sd of 5 in one statement.
> 2. Can you calculate the mean of r1? What about the standard deviation?

All R objects have two basic characteristics, they have a length (discussed previously) and a **mode**.

```
mode(r1)
```

```
## [1] "numeric"
```

Data are traditionally stored in four types of modes. These are "numeric," "character," "complex," or "logical" (TRUE vs FALSE). There are increasingly different types of modes in more specialized packages, but for now we'll just focus on these four. Some types of objects can be switched between modes, but not others, and it is useful to think about when and why you might want to do this. For example, we can change numeric objects into characters.

```
#Let's make an object and call it something simple
B <- 5
B
```

```
## [1] 5
```

```
#What is the mode of the object?
mode(B)
```

```
## [1] "numeric"
```

```
#Let's change the mode
B <- as.character(B)
B
```

```
## [1] "5"
```

```
#What is the new mode of the object?
mode(B)
```

```
## [1] "character"
```

What did we do there? We first assigned the value 5 to the object **B**, and we verified that the mode was indeed numeric. Then, we used the function **as.character()** to assign a new mode to object **B**. Now, **B** is not the value 5, but rather the text representation of 5. Hence, why it has quotes around it.

This all probably seems very basic, but it is useful to know the mode of an object or a vector for several reasons. First, when you import a dataset R tries to be smart about determining the mode. *A vector can only contain one mode*, and R is only as smart as you are. In other words, you can't have both numbers and letters in a single vector. Most data that we are accustomed to working with will be in a structure called a **data frame**, which is essentially just a collection of vectors formed into columns. Since a vector can only have one mode, if you import a data frame from Excel (for example) and you have both numbers and text in a single column, R will automatically make it all characters (since you can't really go the other way and make a numerical representation of a character). We will discuss data frames in more detail a little later in this chapter.

There is one special set of characters that can be used in R without changing the mode of a numeric vector: **NA**, which means "Not Available." Use NA to represent missing data. It may go without saying, but it is absolutely not acceptable to use zeros (0) to signify missing data, since zero is treated as an observation in R. Therefore, you should code NAs directly into your data set when data are missing. That being said, NAs can cause problems with some functions if not properly accounted for in the function. For example, an NA in a vector will cause the **mean()** function to return a value of NA instead of the mean you are looking for.

```
#Make a new vector of 101 values,
#100 of which are numbers and 1 is an NA
r2 <- c(1:100,NA)
#Calculate the mean
mean(r2)
```

```
## [1] NA
```

Taking the mean of **r2** returns an NA, which is probably not what you wanted to have happen. For some functions, you need to either remove the NAs first or specify that NAs should be ignored. For the **mean()** function, that can be done by adding one extra argument. Adding the argument "*na.rm=TRUE*" is like saying "Remove NAs? Yes please."

```
#Calculate the mean, but remove the NAs
mean(r2, na.rm=TRUE)
```

```
## [1] 50.5
```

1.8 WHAT YOUR DATA SHOULD LOOK LIKE BEFORE LOADING INTO R

Spreadsheets (MS Excel, Google Sheets, etc.) are very useful for entering data, but not necessarily for analyzing data. Their flexible nature enables the user to enter all sorts of different pages with a variety of notes and a way to store those data. You can do things like color code individual cells or columns. In my experience, people are often terrible at organizing their data in a way that makes it useful for analysis. Most of the time, folks view their spreadsheets as a simple place to dump information. Your Excel file is not your scrapbook! For example, look at Figure 1.3.

On one level, this might seem like an intuitive way to enter our data. We can clearly see that Susan measured the animals in Block 1 and Darren measured the animals in Block 2. We can see that each person measured two tanks per block, and they measured 4 animals in each tank. Great right? But if you look closer, you can see that Susan called the four animals in each tank the same things (tadpoles 1–4), whereas Darren gave them unique IDs (tadpoles 1–8). Susan used lowercase letters to abbreviate snout-vent length

Block 1	Measured by Susan			Block 2	Measured by Darren		
Tank 1	svl	Tail		Tank 3	None	Tail	
Tadpole 1	6	12		Tadpole 1	7		16
Tadpole 2	5	5		Tadpole 2	6		4
Tadpole 3	8	5		Tadpole 3	9	None	
Tadpole 4	12	0.9		Tadpole 4	4		3
Tank 2	SVL	Tail Length		Tank 4	None	Tail	
Tadpole 1	6			Tadpole 5	7		16
Tadpole 2	5			Tadpole6	5		4
Tadpole 3	8			Tadpole 7	7	None	
Tadpole 4	12			Tadpole 8	9		3

Figure 1.3 An example of the wrong way to organize your data.

(SVL) for tank 1 but capitalized it for tank 2. Darren forgot to include a space in between "Tadpole" and "6." R will treat typos like these as separate and independent, causing problems. Evidently Susan didn't measure the tails in tank 2 at all and just left the cells blank. Darren had two tadpoles without measurable tails and wrote the word "none" in each cell. All of these sorts of things would make the data impossible to analyze.

A much better way to organize these data is shown in Figure 1.4. What you want to aim for is one observation per row, which places data into a relatively long format. In this version, you have a separate column for each type of measurement you have taken and a separate row for each individual that has been measured. Each individual is given a unique identifier. NAs are used in place of any missing data. It might seem weird to have things repeated on many lines, such as the name of the measurer (Susan vs Darren) but you want each row to have all the information necessary to identify it.

	A	B	C	D	E	F
1	Tank	Block	Measurer	Tadpole	SVL	Tail Length
2	1	1	Susan	1	6	12
3	1	1	Susan	2	5	5
4	1	1	Susan	3	8	5
5	1	1	Susan	4	12	0.9
6	2	1	Susan	5	6	NA
7	2	1	Susan	6	5	NA
8	2	1	Susan	7	8	NA
9	2	1	Susan	8	12	NA
10	3	2	Darren	9	7	16
11	3	2	Darren	10	6	4
12	3	2	Darren	11	9	0
13	3	2	Darren	12	4	3
14	4	2	Darren	13	7	16
15	4	2	Darren	14	5	4
16	4	2	Darren	15	7	0
17	4	2	Darren	16	9	3
18						

Figure 1.4 An example of a much better way to organize your data.

Table 1.1 Some various types of data objects available in R

Object	Modes	Several modes possible?
vector	numeric, character, logical, or complex	No
matrix	numeric, character, logical, or complex	No
data frame	numeric, character, logical, or complex	Yes
list	numeric, character, logical, complex ...	Yes

1.9 UNDERSTANDING VARIOUS TYPES OF OBJECTS IN R

There are a number of different types of objects in R, and it is important to understand how each of these work (Table 1.1). There are certainly more types of objects than these, but these are the foundational objects you need to understand for now. For each of these types of objects (and most anything in R), we can ask R what kind of object it is by using the **str()** function (short for "structure").

1.9.1 Vector

A vector has only a single dimension, it is a sequence of elements that are all the same type. The length of the vector is defined by the number of elements in the vector. All of these must be the same mode (hopefully you remember what the mode is from just a few pages ago!).

```
str(r1)
```

```
##   num [1:100] 6.87 4.38 4.38 6.58 8.62 ...
```

The structure of object **r1** tells us it is a numeric vector with 100 elements, and it gives us the first five elements. We can also ask R directly if **r1** is a vector.

```
is.vector(r1)
```

```
## [1] TRUE
```

1.9.2 Matrix

A matrix is essentially a vector that has been given an additional attribute, which is just where to wrap around to create multiple rows or columns. Thus, it's a vector that has a 2-dimensional structure. Since it is basically just a fancy vector, all the elements in a matrix still need to be of the same mode (e.g. "numeric," "logical," etc.).

To create a matrix, we can specify the data to start off with, plus the number of rows and columns and if the data should be wrapped based on rows or columns (with the "*byrow=*" argument). Note that there is a "*bycol=*" argument which does the opposite of "*byrow=*." Also note that by saying you don't want to wrap by row ("*byrow=FALSE*") you are doing the exact same thing as saying "*bycol=TRUE*."

```
#Make a matrix and wrap based on rows
mat1<-matrix(data=1:25,ncol=5,nrow=5,byrow=TRUE)
mat1
```

```
##       [,1] [,2] [,3] [,4] [,5]
## [1,]    1    2    3    4    5
## [2,]    6    7    8    9   10
## [3,]   11   12   13   14   15
## [4,]   16   17   18   19   20
## [5,]   21   22   23   24   25
```

```
#Make a matrix and don't wrap based on rows
mat2<-matrix(data=1:25,ncol=5,nrow=5,byrow=FALSE)
mat2
```

```
##       [,1] [,2] [,3] [,4] [,5]
## [1,]    1    6   11   16   21
```

```
## [2,]    2    7   12    17    22
## [3,]    3    8   13    18    23
## [4,]    4    9   14    19    24
## [5,]    5   10   15    20    25
```

1.9.3 Data frame

A data frame is probably the most useful and most used of the objects we will discuss in this book. I know I said earlier that vectors are the most important, and they are, but a data frame is essentially a table composed of one or more vectors. Thus, if you can understand vectors you can understand data frames. All of the vectors in a data frame have to have the same length (which is important), but the data in those vectors can be different modes (which is also important). We will learn how to read in data in Chapter 3. For now, let's create a data frame from scratch, which also provides an opportunity to introduce some useful basic functions.

Let's make a vector of values and a vector of names (you might imagine they are treatment groups, for example). We can use the function **rep()** to repeat something as many times as we want. For example, let's imagine we have two treatments each with 20 individuals, and that the average value of whatever we have measured is 5 for one group and 10 for the other group. In the code below we have nested several functions to achieve what we want to do. In the first line, we use the **c()** function to concatenate the words *Group.A* and *Group.B* into a single vector, then use the **rep()** function to repeat each value in that vector 20 times. What happens if you replace the argument "*each=*" with "*times=?*" In the second line, we concatenate together two vectors that are each 20 numbers long. In the third line we use the function **data.frame()** to create the new data frame from our two vectors and store it as an object called **df1**.

```
#First, make a vector of names for the groups
treatment<-rep(c("Group.A","Group.B"), each=20)
#Next, make a vector of hypothetical values for each group
values<-c(rnorm(20,5,1),rnorm(20,10,1))
#Combine the two vectors into a data frame
```

```
df1<-data.frame(treatment,values)
df1
```

```
##    treatment    values
## 1    Group.A  5.326256
## 2    Group.A  6.271354
## 3    Group.A  4.083671
## 4    Group.A  4.679081
## 5    Group.A  6.435922
## 6    Group.A  6.302879
## 7    Group.A  6.165815
## 8    Group.A  4.231960
## 9    Group.A  5.770652
## 10   Group.A  5.036748
## 11   Group.A  5.582768
## 12   Group.A  3.128528
## 13   Group.A  6.783060
## 14   Group.A  1.886150
## 15   Group.A  3.889663
## 16   Group.A  5.703414
## 17   Group.A  5.755549
## 18   Group.A  4.797857
## 19   Group.A  4.269279
## 20   Group.A  5.658149
## 21   Group.B 11.464389
## 22   Group.B  7.974523
## 23   Group.B  9.383801
## 24   Group.B 10.298131
## 25   Group.B 11.174720
## 26   Group.B 12.393812
## 27   Group.B 10.646669
## 28   Group.B  8.601714
## 29   Group.B  9.425939
## 30   Group.B 10.331108
## 31   Group.B 10.481751
## 32   Group.B 10.850636
## 33   Group.B 11.421954
```

```
## 34     Group.B  9.775505
## 35     Group.B  9.766465
## 36     Group.B  8.471132
## 37     Group.B  8.861892
## 38     Group.B  8.693411
## 39     Group.B  9.416204
## 40     Group.B 10.660106
```

As discussed previously, there is nothing special about the names chosen here. We could have called our vectors whatever we wanted, but the names **treatment** and **values** are sensible names to use for our purposes. Note that making the data frame **df1** from the two vectors only works because we had already created the objects **treatment** and **values**. If you called your vectors "vec1" and "vec2," you would need to modify the code where you create the data frame accordingly.

Earlier we use the function **str()** to look at the *structure* of a vector. The same function works for getting a quick look at a data frame. I think you will find that **str()** is one of the most useful functions there is. It quickly tells you what sort of data are found in each column in your data frame, as well as the size of the data frame. The number of "*obs.*" is the number of rows in the data frame and the number of "*variables*" is the number of columns.

```
str(df1)
```

```
## 'data.frame':    40 obs. of  2 variables:
##  $ treatment: chr  "Group.A" "Group.A" "Group.A" "Group.A" ...
##  $ values   : num  5.33 6.27 4.08 4.68 6.44 ...
```

For obvious reasons, it can sometimes be useful to change the names of columns in a data frame. There are two easy ways to do this. First, we can reassign the column names of the data frame using the **names()** function. Note that below I've used a function called **head()**, which just shows the first 6 rows of the data frame (as opposed to the whole thing).

```
#What are the current column names?
names(df1)
```

```
## [1] "treatment" "values"
```

```
#Assign new names to the columns
names(df1)<-c("Groups","Data")
#Look at the top of the data frame
head(df1)
```

```
##     Groups      Data
## 1 Group.A 5.326256
## 2 Group.A 6.271354
## 3 Group.A 4.083671
## 4 Group.A 4.679081
## 5 Group.A 6.435922
## 6 Group.A 6.302879
```

You can also assign column names when you first create the data frame.

```
df1<-data.frame("Groups" = treatment, "Data" = values)
head(df1)
```

```
##     Groups      Data
## 1 Group.A 5.326256
## 2 Group.A 6.271354
## 3 Group.A 4.083671
## 4 Group.A 4.679081
## 5 Group.A 6.435922
## 6 Group.A 6.302879
```

Data frames can be navigated with square brackets just like vectors, but now we have two dimensions present. Thus, we will now put two numbers inside the brackets, representing the rows and columns (in that order), separated by a comma. Thus, you read it as "[rows,columns]." If we wanted to access the element in the 3rd row and 2nd column of df1, we would type the following.

```
df1[3,2]
```

```
## [1] 4.083671
```

We can also index entire rows or entire columns if we leave one side of the comma blank. Thus, to index all of row 7, you would type the following.

```
df1[7,]
```

```
##     Groups     Data
## 7 Group.A 6.165815
```

The same logic applies of course to indexing entire columns. Just leave the row side blank and indicate which column you want to look at. Go ahead, try it out for yourself.

It can also be very useful to *summarize* your data frame. Whereas **str()** tells about the structure of the data frame, **summary()** will tell us basic information about each column in our data frame. What's better still is that R tells us different information about columns with different types of data. For a column of numeric values like our column **Data**, R will tell us the min and max values, the mean and median, and interquartile range.

```
summary(df1)
```

```
##     Groups                Data
##  Length:40          Min.   : 1.886
##  Class :character   1st Qu.: 5.519
##  Mode  :character   Median : 7.379
##                     Mean   : 7.546
##                     3rd Qu.: 9.906
##                     Max.   :12.394
```

Another, extremely useful, way to access information in a data frame is with the dollar sign operator, **$**. The dollar sign allows you to access a column by its name. Just type the name of the data frame, then the **$**, then the column name. For example, if we want to look at the **Data** column, you can type the following, which returns the vector of values in the column.

```
df1$Data
```

```
##  [1]  5.058114  6.539179  4.859611  5.601500  5.666716  6.521534
##  [7]  3.233818  4.695411  5.287843  6.174719  4.391372  3.264850
## [13]  3.930722  3.319717  5.652374  5.926697  6.624910  6.571445
## [19]  5.272495  5.339407 12.762116  8.807038  9.403908 10.410916
## [25]  9.035601  9.084326 10.246616  9.199376  7.289678 10.629323
## [31] 11.331721 11.579284  8.567594  8.946644  9.037035 10.151584
## [37] 10.707575 10.125538 11.566409  8.706914
```

Since that represents the vector of values, you can also index it. For example, if we want the 10th through 20th objects in the **Data** column of **df1**, we can do the following. Note that since we are indexing a vector, we are back to a one-dimensional sequence of values and our indexing does not have a comma to separate rows and columns.

```
df1$Data[10:20]
```

```
##  [1] 6.174719 4.391372 3.264850 3.930722 3.319717 5.652374
##  [7] 5.926697 6.624910 6.571445 5.272495 5.339407
```

1.9.4 Lists

Finally, lists are potentially complex objects that can contain any type of object (vectors, matrices, data frames, even other lists). Lists are often generated by the analyses we will cover in later chapters (e.g., the output for a linear model is a list).

Let's say we want to create a list that has both **df1** and **mat2** in it. We can do that with the following code:

```
list1<-list(df1,mat2)
list1
```

```
## [[1]]
##      Groups      Data
## 1  Group.A  5.326256
## 2  Group.A  6.271354
## 3  Group.A  4.083671
## 4  Group.A  4.679081
## 5  Group.A  6.435922
## 6  Group.A  6.302879
## 7  Group.A  6.165815
## 8  Group.A  4.231960
## 9  Group.A  5.770652
## 10 Group.A  5.036748
## 11 Group.A  5.582768
## 12 Group.A  3.128528
## 13 Group.A  6.783060
## 14 Group.A  1.886150
## 15 Group.A  3.889663
## 16 Group.A  5.703414
## 17 Group.A  5.755549
## 18 Group.A  4.797857
## 19 Group.A  4.269279
## 20 Group.A  5.658149
## 21 Group.B 11.464389
## 22 Group.B  7.974523
## 23 Group.B  9.383801
## 24 Group.B 10.298131
## 25 Group.B 11.174720
## 26 Group.B 12.393812
## 27 Group.B 10.646669
## 28 Group.B  8.601714
## 29 Group.B  9.425939
## 30 Group.B 10.331108
## 31 Group.B 10.481751
## 32 Group.B 10.850636
## 33 Group.B 11.421954
## 34 Group.B  9.775505
## 35 Group.B  9.766465
```

```
## 36 Group.B  8.471132
## 37 Group.B  8.861892
## 38 Group.B  8.693411
## 39 Group.B  9.416204
## 40 Group.B 10.660106
##
## [[2]]
##      [,1] [,2] [,3] [,4] [,5]
## [1,]    1    6   11   16   21
## [2,]    2    7   12   17   22
## [3,]    3    8   13   18   23
## [4,]    4    9   14   19   24
## [5,]    5   10   15   20   25
```

Did R keep those both the same? How can we tell? Recall that we can look
at the structure of any object with **str()**.

```
str(list1)
```

```
## List of 2
##  $ :'data.frame':    40 obs. of  2 variables:
##   ..$ Groups: chr [1:40] "Group.A" "Group.A" "Group.A" "Group.A" ...
##   ..$ Data  : num [1:40] 5.33 6.27 4.08 4.68 6.44 ...
##  $ : int [1:5, 1:5] 1 2 3 4 5 6 7 8 9 10 ...
```

If we want to access one of the objects in a list, we can once again use
indexing, but here we have double square brackets "[[]]" to indicate which
object in the list we are referring to.

```
list1[[2]]
```

```
##      [,1] [,2] [,3] [,4] [,5]
## [1,]    1    6   11   16   21
## [2,]    2    7   12   17   22
## [3,]    3    8   13   18   23
## [4,]    4    9   14   19   24
## [5,]    5   10   15   20   25
```

What would we do if we wanted to access the value that is in the 5th row, 2nd column of the **df1** part of the list?

```
list1[[1]][5,2]
```

```
## [1] 6.435922
```

What you just read or typed along with may seem trivial, but it is extremely crucial! R works sequentially left-to-right in most commands and it allows you to compound together multiple commands one-after-another. Thus, typing **list1[[1]]** is *exactly the same* as the typing **df1**. It is representative of the data frame, which means that we can tag indexing commands on to the end of it in order to access things inside the data frame. Don't worry if that does not make perfect sense right now, it will sink in over time and become clear as we work through more examples.

We can also name the objects in a list if we want.

```
names(list1)<-c("my.data.frame","my.matrix")
str(list1)
```

```
## List of 2
##  $ my.data.frame:'data.frame':    40 obs. of  2 variables:
##   ..$ Groups: chr [1:40] "Group.A" "Group.A" "Group.A" "Group.A" ...
##   ..$ Data  : num [1:40] 5.33 6.27 4.08 4.68 6.44 ...
##  $ my.matrix    : int [1:5, 1:5] 1 2 3 4 5 6 7 8 9 10 ...
```

Once the object has names, you can use the name to pull out the specific object alone with the dollar sign to signify the name, just like we did with previous data frames.

```
list1$my.matrix
```

```
##      [,1] [,2] [,3] [,4] [,5]
## [1,]    1    6   11   16   21
## [2,]    2    7   12   17   22
## [3,]    3    8   13   18   23
## [4,]    4    9   14   19   24
## [5,]    5   10   15   20   25
```

1.10 A LITANY OF USEFUL FUNCTIONS

It is impossible to try and provide a comprehensive list of the functions you could use in R. Here is a brief list of my favorite and/or most used simple commands.

1.10.1 Functions for examining data frames or other objects

1. **ncol()** - Tells you the number of columns in a data frame.
2. **nrow()** - Tells you the number of rows in a data frame.
3. **head()** - Shows you the top 6 rows of a data frame. Note that you can specify more rows if you want.
4. **tail()** - Shows you the last 6 rows of a data frame. Note that you can specify more rows if you want.
5. **names()** - Shows you the column names and allows you to change them.
6. **colSums()** - Calculate the sums of multiple columns in a data frame.
7. **colMeans()** - Calculate the means of multiple columns in a data frame.

1.10.2 General functions for creating or manipulating objects

1. **seq()** - Create a sequence of numbers. You can specify the start and end values and how many values should be in between them.
2. **rep()** - Repeat something a set number of times.
3. **trunc()** - Truncate a number to just the integer.
4. **round()** - Round a number to a set number of decimal places.
5. **rbind()** - Bind two vectors together as rows.
6. **cbind()** - Bind two vectors together as columns.
7. **paste()** - Stick two bits of text together. You can specify what character, if any, you want to use to separate them.

1.10.3 Mathematic functions

1. **mean()** - Calculate the mean of a vector of numbers.
2. **sum()** - Calculate the sum of a vector of numbers.

3. **max()** - Find the maximum value in a vector.

4. **min()** - Find the minimum value in a vector.

5. **length()** - Tells you how long a vector is.

6. **range()** - Tells you the max and min values of a vector.

7. **sd()** - Calculate the standard deviation of a vector.

8. **abs()** - Take the absolute value of a vector.

┤ **Box 1.9 - Take-home points** ├

- Always work from a script and annotate your code. Creating a clear and intelligible code file is extremely important, both for yourself and for others who you will share your code with eventually.
- R works by creating objects. Use the assign function (the little arrow: **<-**) to create named objects and assign values or characters to them.
- Functions are pieces of code that do specific tasks. Functions take **arguments** which allow you to customize what the function is doing to your needs. Importantly, functions often have default values for many of their arguments and it is useful to know what those defaults are.
- Be smart about how you set up your data file before bringing it into R. Remember, your spreadsheet is not your scrapbook.
- Vectors and data frames are the most essential and important basic units in R. Understanding how to navigate and manipulate them is key.

1.11 ASSIGNMENT!

Here are some things to do on your own. Remember you can find example answers and code at https://github.com/jtouchon/Applied-Statistics-with-R.

1. Create a data frame that contains at least one factor with three categories and at least three columns of made up numeric data. Thus, it should have at least four columns. Make sure the columns have meaningful names. Have at least 10 rows per categories (i.e., at least 30 rows long).

2. Once you have a data frame, plot a histogram of your numeric variables. Try to dress it up by adding color, changing the limits

of the axes, and adding a main title. How will you figure out how to do that? Look at the help file for **hist()**.

3. Use the **plot()** function to try and plot your numeric and categorical variables. What happens when you give R different types of data to plot?

Before You Begin (aka Thoughts on Proper Data Analysis)

2 Before You Begin (aka Thoughts on Proper Data Analysis)

2.1 OVERVIEW

Before we embark on the journey that is learning R and how to use it to analyze your data and make fantastic figures, it is useful to stop and think a little bit about best practices for data analysis.

2.2 BASIC PRINCIPLES OF EXPERIMENTAL DESIGN

If there are three words to remember when thinking about experimental design, they are **balance**, **randomization**, and **replication**. In a nutshell, what you are trying to prevent with these three factors is your data being correlated in some way that is unhelpful to your analysis. You are also trying to ensure that your data are independent from one another and that you have enough data to actually determine if your treatments did anything or not (see Boxes 2.1, 2.2, and 2.3).

Applied Statistics with R: A Practical Guide for the Life Sciences. Justin C. Touchon, Oxford University Press (2021). © Justin C. Touchon. DOI: 10.1093/oso/9780198869979.003.0002

Box 2.1 - Balance

As much as is possible, you should aim to have relatively even numbers of individuals in whatever treatments you might have. Unbalanced experimental designs lead to uneven sample sizes, which can cause problems for your analyses.

Box 2.2 - Randomization

You should have individuals randomly allocated into whatever treatment groups you have. Always make sure that you set up your experiment without any knowledge of the possible pre-existing nature of your individuals. For example, if you are doing a growth experiment, don't pick all the biggest individuals to be in one particular treatment, since it will bias your results. Make sure you allocate individuals blindly. This can be accomplished by numbering all of your individuals and then randomizing the numbers into your treatments.

Box 2.3 - Replication

It should be obvious, but the more data you have, the better off you are. However, replication does not just mean having more individuals. Replication refers to actually repeatedly setting up the experiment so that you have independent data points to measure. For example, imagine you want to test the effect of two different fertilizers on plant growth and you have 100 seeds to start with. There are many different ways you could actually set up your experiment. You could set up 50 seeds each in two pots of soil, one for each treatment, but that would give you no replication. It might at first seem like you are growing 100 plants which is a lot! But from a statistical perspective, you have a sample size of 1 in each treatment. This is known as *pseudoreplication*, a term first coined by Stuart Hurlbert in his classic 1984 paper "Pseudoreplication and the design of ecological field experiments." At the other extreme, you could set up 1 seed in each of 100 different pots and provide 50 with one fertilizer and 50 with the other. In each case you are growing the same number of plants, but in the second case each plant can be treated as independent from each other plant. The first case has no replication whatsoever whereas the second case is fully replicated. A third option would be to have 10 seeds growing in each of 10 pots, with 5 pots each getting each type of fertilizer. Again, this is the same number of seeds but now you have grouped data with some degree of replication. We will explore the idea of grouping later in this chapter.

Obviously, it is not always possible to control these aspects of your data, particularly if you have observational data (as compared to a controlled experiment which you design and run yourself). But, even in the case of observational studies, these principles are important to keep in mind and consider.

2.3 BLOCKED EXPERIMENTAL DESIGNS

Consider the four experimental setups shown in Figure 2.1. Imagine that we are now testing the effects of four fertilizers on plant growth (labelled A, B, C, and D), each with 12 individuals. The experiment is conducted in four separate "blocks." *What is a block?* It could be many things. Maybe it is a physical way of setting up the experiment, for example, four shelves in an incubator that contain the experimental units or four rooms that contain the cages our individuals live in. Maybe, due to space or time limitations, only 12 individuals can be tested or measured at a time, and thus the experiment has to be run four separate times. Each of these can be considered a "block" so you can hopefully imagine how this idea relates to your own research. Blocks are only important to consider if there is some systematic difference among them.

In the first example, the four treatments will be perfectly correlated with the four blocks. Thus, if we imagine a significant difference is detected in one treatment, there is no way to know if it is because of the experimental treatment or if there was something else going on in that block (or room, or time point, or whatever you want to imagine that block represents). Once again, this is an example of pseudoreplication because it seems like we have a large sample size but in reality, we have a sample size of N=1 in each of our treatments. Despite growing 48 different plants, this design is unreplicated.

The second example is a fully randomized design, where the four treatments are allocated across the four blocks completely at random. The third example is a fully balanced design, where each of the four treatments is assigned to each block in the same manner. Each of these setups has its own advantages and disadvantages.

The fully randomized design is good, and in theory should lead to the highest degree of replication, with all experimental units being truly independent. In reality it can actually work against the principle of balance, since some treatments might end up overrepresented in some blocks and underrepresented in others (e.g., in Figure 2.1, there are six individuals from treatment B in Block 2, but only one in Block 3). In the extreme, leaving your entire setup to random chance could lead to a horribly

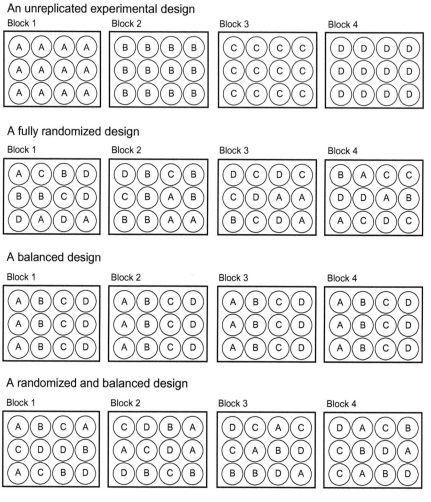

Figure 2.1 Examples of four different possible experimental designs where we are testing the effects of four different fertilizers (labelled simply A, B, C, and D) on plant growth. The experiment is executed in four blocks, which could be considered the physical arrangement of the experimental units or the time course when the experiment is conducted, amongst other possibilities.

unbalanced and biased design, but this would be very rare. In general, fully randomizing your experiment is a very good idea!

Similarly, the balanced design ensures that we have even representation of our four treatments in each block but it is still prone to problems since the physical setup of each block is identical. Imagine that each block is a shelf containing the experimental units. Maybe there is more light on one side of the shelving unit and you get a significant difference in treatment A. Maybe it was just because they were closer to the light? You would not be able to rule that out, and that sure would be unfortunate.

Along with a fully randomized design, an excellent experimental design is the fourth one, where individuals from each replicate are allocated to each block in a balanced manner, but then individuals are randomly set up within each block. This design ensures that you have even representation of your treatments across the blocks, which is important for your data analysis, but also prevents any unwanted correlations due to the physical or temporal way that the experiment is run. If there are strong temporal or spatial components to your blocks, this is likely the best design because it will allow you to more fully account for differences across blocks. If this is the case for your experiment, you will definitely want to read Chapter 8 about mixed effects models.

2.4 YOU CAN (AND SHOULD) PLAN YOUR ANALYSES BEFORE YOU HAVE THE DATA!

In addition to the aspects of experimental design described previously, the other most important thing to do is to have a *clear idea of your predictor and response variables* before you even start the experiment. Before you ever put a mouse in a testing box or a seed in growth chamber, you should identify what it is you are going to measure. Hopefully, if you know your study system pretty well or perhaps have some preliminary data, you can estimate what the data are going look like which will allow you to think about and plan for what type of analyses you will do. Maybe that sounds like wishful thinking, but this whole book is about the importance of knowing what your data look like, so don't worry—you'll get there!

Sir Ronald Fisher, one of the founders of modern statistics, offered one of the best statements about this issue in 1938.

> To consult the statistician after an experiment is finished is often merely to ask him to conduct a *postmortem* examination. He can perhaps say what the experiment died of Ronald Fisher

Fisher's point is that because your experimental design directly effects your data analysis, you should think about your analysis up front when planning the experiment.

Box 2.4 - How to not plan an experiment

This is a pretty embarrassing story for someone writing a statistics book to tell, but hopefully it will serve as an instructive cautionary tale.

I started graduate school in the fall of 2002 and was excited to start doing experiments and collecting data. I had not taken a statistics course as an undergraduate and I figured that whatever experiments I could dream up, there would be some way to analyze the data. Get in the field, collect the data, then figure out what to do with it. That was my plan.

I first travelled to Panama to do field work in June 2003 and would continue to spend 3–6 months there each year until I finished my PhD. A few years in, I got a copy of a statistics program that we can leave unnamed (let's just call it a Statistical Package for Social Sciences). After taking a statistics course where I spent a semester learning how to calculate an ANOVA by hand, I was still unprepared to analyze my data and was left trying to figure it all out on my own. It turned out that the program I was trying to use *simply could not do the analysis I needed*.

Luckily for me, a generous and wonderful postdoc in my lab named James Vonesh (now a professor at Virginia Commonwealth University) introduced me to a little known but free program called R, which could, as he explained, do pretty much anything one could imagine. It wasn't always easy, and you had to invest some time to learn your stuff, but it could be done and would be worth it.

The moral of the story is that you should not be like me, and you should think about your experimental design and how it relates to how you will analyze your data before you get started. It will be worth it, and will save you huge headaches in the end, and you won't have to rely on a generous postdoc to dig you out of a hole!

2.5 BEST PRACTICES FOR DATA ANALYSIS

Once you have started analyzing your data, how should you actually go about the process? By that I don't mean "sit at your computer and type code into R," but what is the best way to think about running your analyses and coming to conclusions about your data?

The most important rule of the road is to have clear hypotheses at the outset of analyzing your data. You ran the experiment (presumably) and so you should have an idea of what questions you are trying to answer. It is important that you do not just try to measure everything under the sun and then look for possible relationships between your variables. If you measure enough things, you are more likely to find a statistically significant relationship that may or may not be meaningful. This is a process known as *p-hacking* or *data dredging*. Looking for patterns in your data is certainly an okay thing to do, but you should avoid testing every possible predictor you can imagine in the hope of finding *something* significant that will make your data more publishable. This is particularly true when you have many possible predictor variables.

A second dubious practice which is to be avoided at all costs is *HARKing*, which stands for "hypothesizing after the results are known." HARKing basically goes like this: 1) You start with a hypothesis that you want to test; 2) you test it and find a non-significant effect; 3) while doing your series of analyses you come across some other variable which does have a significant explanatory effect; and 4), you omit your original hypothesis and concoct a new hypothesis which fits the results of the study. Essentially, you are coming up with a prediction after you know the answer. It is certainly important to realize that you might learn something new when you analyze your data. Perhaps the unforeseen result causes you to see a new explanatory relationship which you had not considered before. That's fine, but it is important to be transparent about your original hypothesis versus the new hypothesis. Moreover, you should not be testing all sorts of predictors that may or may not seem important and then trying to come up with explanatory hypotheses after the fact.

The last piece of advice regarding best practices is *don't obsess over statistical significance*. It has been said many times, but it bears repeating, that the 0.05 cutoff that is traditionally used to demarcate "significance" is a completely arbitrary line in the sand. Why isn't it 0.01 or 0.10? Obviously, the idea of statistical significance is still an oft-used metric in many fields, and we will certainly talk about significance in this book, but don't obsess over it. It isn't the be-all and end-all of your research endeavors, because lots of things affect your ability to detect "significance" beyond the validity of your hypotheses, the magnitude of biological affects you measure, or the strength of your data. The more data you have, the greater your ability to detect very small affects. Similarly, with very little data, you will have a hard time detecting even moderate affects. The more variable your data are from individual to individual, the harder a time you will have detecting significance. So, focus your energy on understanding and estimating the explanatory power of your predictor variables. What you want to know is if there is a relationship between your predictor and response variables and how confident you can be in any relationship you've tried to quantify, so try to keep that in your mind at all times while analyzing your data.

2.6 HOW TO DECIDE BETWEEN COMPETING ANALYSES

Something we will explore later in this book with practical, real-world examples is how to choose amongst different possible ways to analyze your data. However, it is still useful to think about this issue up front, because it is an important one. Many times, you will be faced with different possible ways to analyze your data and different methods might have certain pros and cons.

At the end of the day, your job is to pick the statistical analysis that does the best job of representing and interpreting your data. Getting you to the point where you know *how* to evaluate your data and make that decision is one of the major goals of this book. Sometimes there will be two different types of models that may be essentially equivalent. They may fit equally well, and they may give you very similar answers about the significance of your predictors. Sometimes you might have to choose amongst competing models that give you different answers about the supposed significance (or not) of your predictors. How do you choose which model to go with?

You should be completely neutral about the idea of "significance" and only consider if the model is doing a good job and is appropriate for the data. As we will see in Chapter 7, there are times when you might have a model which might seem appropriate and gives you a highly significant result, but which actually fits very poorly. The opposite can be true as well, where you choose a model that fits poorly and gives you an underwhelming sense of the importance of your predictors. At the end of the day, you need to be able to justify your choice of model to yourself, your reviewers, your supervisors, your colleagues, and anyone else that might look at your analysis. You need to be able to stand behind your decision and explain why you chose one analysis over another. If you can do that, you'll be in good shape.

2.7 DATA ARE DATA ARE DATA

The data that we will work within this book come from a study I conducted as a postdoc working in the rainforests of Panama at the Smithsonian Tropical Research Institute. I have studied frogs and their eggs and tadpoles since I began my PhD in 2002. I love frogs and their eggs and tadpoles. I think they are the greatest. But, maybe you don't, and maybe you are thinking to yourself: "How do these weird tadpole data relate to me and my analyses? I'm never going to study tadpoles in Panama!" While that might be true, I think that data are data are data, and the spatially nested design of the study we will work through is extremely similar to many other study systems (Figure 2.2).

The shorthand name of the dataset we will work with is "RxP," which stands for "Resource X Predation," which are the two main things we were manipulating in the study (Chapter 3 will describe the study in more detail). In the RxP dataset, we have individuals that are nested (or grouped, if you will) within tanks, and those tanks are themselves nested within blocks. This is the same as, for example, having multiple mice housed together in a single cage, and then having cages grouped together on different shelves. Or, you might imagine multiple fruit flies measured per vial, and vials are grouped together based on the date they were set up. Similarly, we can think about having multiple genetic strains (i.e., genotypes, families, etc.) that cluster our data into discrete groups. This sort of nested design can also be used to think about repeated measures data, where individuals are measured or sampled repeatedly over time.

Beyond the idea of physically clustered data, we can just think about the tanks as different replicates of an experiment, which is no different from any other experiment. The response variables (also called the "dependent variables") that were measured during the RxP experiment—things like number of days until metamorphosis or size at metamorphosis—are no different than any other continuous variable that might be measured, be it the length of time a rat freezes after seeing a light flash or the number of pecks a pigeon makes before correctly opening a lever or the number of new leaves produced after adding fertilizer to a growing plant. The predictor variables (also called the "independent variables") are generally categorical in their nature (i.e., they have several discrete categories), and could easily be replaced by drug treatment or maternal enrichment environment or fertilization regime. As I said earlier, data are data are data. See Figure 2.2 for a visual depiction of the similarities in data structure.

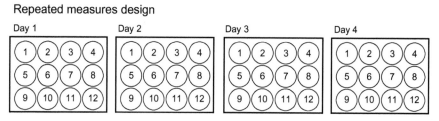

Figure 2.2 The RxP dataset used in this book is a spatially nested design, where individuals are nested within tanks, and each tank is nested within a larger block. While these data are ecological in nature, they are no different than cells that might be nested within culture plates or mice that might be nested within cages, which are themselves nested on different shelves. This sort of design can also be used to study repeated measures data, where individuals are tested over multiple time points, or to control for things like genetic relatedness (i.e., families). These last two designs will be discussed in Chapter 8 on mixed effects models.

Box 2.5 - Take-home message

- The three most important things to keep in mind are balance, randomization, and replication. In many cases, a combination of balance and randomization will be the best approach.
- You should do your best to plan your analyses before you ever conduct your experiment. This is informative in many ways, but most of all because thinking about how you will analyze your data will influence the way you design your experiment.
- Once you have collected your data, avoid dubious practices such as p-hacking or HARKing. Instead, try to evaluate just your pre-existing hypotheses and be as neutral as possible when evaluating if models are appropriate for explaining your response variables.
- Always remember that data are just data. Regardless of the particular field you are working in, numbers and categories are just that: numbers and categories. The most important thing is that you understand what your data look like, which is what we will begin to cover in Chapter 3.

Exploratory Data Analysis and Data Summarization

3 Exploratory Data Analysis and Data Summarization

The purpose of this chapter is to introduce you to working with and manipulating data in R, exploring data, and plotting. I strongly believe in learning by doing, so let's start doing some things so you can start learning!

3.1 THE RESOURCE-BY-PREDATION DATASET

In this book, we will use data from an experiment I conducted when I was a postdoc at the Smithsonian Tropical Research Institute in Panama in 2010, and which was published in the journal *Ecology* in 2013 (https://www.jstor.org/stable/23436298). The experiment was part of a National Science Foundation (NSF) funded project to Drs. Karen Warkentin (Boston University) and James Vonesh (Virginia Commonwealth University) studying the effects of flexible hatching timing by red-eyed treefrog (*Agalychnis callidryas*) embryos on interactions with predators and food levels and subsequent phenotype development of tadpoles. In order to follow along with the examples in this chapter, and the rest of the

Applied Statistics with R: A Practical Guide for the Life Sciences. Justin C. Touchon, Oxford University Press (2021). © Justin C. Touchon. DOI: 10.1093/oso/9780198869979.003.0003

book, you should download from the Github page for this book (https://github.com/jtouchon/Applied-Statistics-with-R) a .csv file titled "RxP.csv." The data are called by the short name "RxP," which stands for "Resource-by-Predation," which was the nature of the experiment (we were studying the interaction of resources and predators). This brings up a chance to reiterate a small but important point: since R is entirely based on typing commands by hand, you should give your datasets and variables short names so that they are quick and easy to type.

First, let's get a handle on what the data are.

Box 3.1 - The RxP data

- The experiment consisted of 96 400 L mesocosm tanks (i.e., artificial ponds) arrayed in an open field in the northwest corner of a small town in Panama called Gamboa, where the Smithsonian runs a field station for tropical research. We wanted to know how variation in when a frog embryo hatches might affect its development to metamorphosis under various combinations of predators and resources.
- Experimental treatments were as follows:
  - Hatching age: Early (4 days post-oviposition) or Late (6 days post-oviposition)
  - Predators: Control, Nonlethal (caged) dragonfly larvae or lethal (free-swimming) dragonfly larvae
  - Resources: Low (0.75 g) or High (1.5 g) food level, added every 5 days
- The mesocosms were spatially arranged in 8 blocks of 12 tanks each. The three treatments were "fully crossed" meaning that each possible combination of treatments was created (e.g., Tank 1 might be Early hatching, Control predator, Low resources, whereas Tank 2 might be Late hatching, Control predator, Low resources). The replicates were fully balanced across blocks and randomized within each block. Thus, each block consisted of one tank from each of the 12 unique treatment combinations, but within each block, the physical setup of tanks in the field was randomly assigned. Each tank began with 50 tadpoles and the experiment ended when all tadpoles reached metamorphosis or had died.
- Metamorphosis is a process that takes time. This process is generally defined as the time from when a froglet's arms erupt from the body (they first develop under the skin) until when the tail is fully resorbed into the body. The froglet may choose to leave the water early or late during that process. Thus, several measurements were taken when a froglet first left the water and several more when the tail was fully resorbed. At metamorphosis, we measured the following response variables:
  - Age at hatching, both in terms of time since eggs were oviposited, and time since the very first froglet crawled out of the water (defined as Day 1)
  - Snout-vent length at emergence, which is the length from the tip of the froglet's nose to its cloaca

- Tail length at emergence
- Snout–vent length at completion of tail resorption
- Mass at completion of tail resorption
- Number of days needed for each metamorphose to fully resorb the tail
- During the course of the experiment disease broke out in 18 of the mesocosms containing Nonlethal predators and thus those tanks have been removed from the dataset.
- Full citation for the article: Touchon, J.C., McCoy, M.W., Vonesh, J.R., and Warkentin, K.W. 2013. Effects of hatching plasticity carry over through metamorphosis in red-eyed treefrogs. *Ecology* 94(4): 850–60.

3.2 READING IN THE DATA FILE

There are several options for reading in a data file, depending on what type of computer you are using and if you are using RStudio or standard R.

3.2.1 Reading the file using **read.csv()**

If you are loading a data file from somewhere on your computer, you will read it into the active workspace with the command **read.csv()**. If your dataset has categorical variables, you will want to include the argument *stringsAsFactors=T*, which tells the function to automatically make any columns that have character data in them factors. If the data are in your working directory, you would simply do the following (and remember, you should be working in a script window!). Note that you have to assign the data to an object. What happens if you do not?

```
RxP<-read.csv("RxP.csv", stringsAsFactors=T)
```

What is the working directory you ask? It is the directory that R will look in by default. To find out where R is looking, type **getwd()** at the prompt.

```
getwd()
```

```
## [1] "/Users/jutouchon/Desktop/Stats_Book"
```

You can set your working directory with the function **setwd()**, where you would put in the parenthesis a path to a folder on your computer. Make sure to put the path in quotes. Some folks like to set a working directory for each project they have. Others prefer to keep the working directly in a single place and just code the path to certain files. Choose whichever works for you. If your data are not in your working directory, you will need to specify *exactly* where to find the file on your computer.

Box 3.2 - File paths on a Mac vs PC

It isn't always easy to figure out exactly where on your computer a file is located. Luckily, if you use a Mac there is a trick built into R to make it easy. Just drag the icon of the file from a Finder window into your script window and your computer will paste in the exact address of the file for you. Note that this does not work in RStudio, just in regular old R. On a PC, it is not as simple. You can copy the address from the file and paste it into your script window but know that for whatever reason it pastes in the address with the forward slashes (/) in the wrong direction, as backslashes (\), and you will have to change that manually.

Box 3.3 - Using RStudio

Under the File menu, there is an option titled "Import Dataset" and you can choose to import the data from your .csv file. In the current version of RStudio, there are two options for how to import the data. The first uses "base" R, which means is uses **read.csv()** exactly as we did previously. The nice thing here though is that you can navigate to where the file is on your computer just like you would any other file. There is also a box you can select to tell it to treat your categorical data as factors. The second option uses the package *readr* which utilizes the function **read_csv()**, which does not read in the file as a *data frame*, but instead as a format called a *tbl_df*, which is slightly different. This is a newer format preferred by RStudio, but the annoying thing is that the format does not always play nice with some of R's older built-in functions. Additionally, there is no way to tell **read_csv()** to automatically make your character data factors, so you would have to do it manually. Evidently, **read_csv()** is much faster at reading in enormous files.

3.3 DATA EXPLORATION AND ERROR CHECKING

Whenever you start working with a dataset in R, you should first devote substantial time to checking it for errors. Questions you should ask yourself include:

- Did the data import correctly?
- Are the column names correct?
- Are the types of data appropriate? (e.g., factor vs numerical)
- Are the numbers of columns and rows appropriate?
- Are there typos?

If, for example, a column that is supposed to be numerical shows up as a factor, that likely indicates a typo where you accidentally have text in place of a number (remember, each column in a data frame is a vector, and vectors can only have one mode, so a vector with both numbers and characters is treated as if it is all characters). Similarly, if you have a factor that should have 3 categories, but imports with 4, you likely have a typo (e.g., "predator" vs "predtaor"), and the misspelled version is showing up as a separate category. These sorts of mistakes are very common!

Because this dataset has been thoroughly examined (very thoroughly!), these types of errors are not present. However, you might want to change the names of columns or remove outliers, which we will cover in the subsequent sections.

3.3.1 Data structure
Begin by examining the structure of the data frame with the function **str()**.

```
str(RxP)
```

```
## 'data.frame':     2502 obs. of  14 variables:
##  $ Ind            : int  1 2 3 4 5 6 7 8 9 10 ...
##  $ Block          : int  5 5 5 5 5 5 5 2 2 1 ...
##  $ Tank           : int  7 4 4 7 10 4 4 5 4 1 ...
##  $ Tank.Unique    : int  55 52 52 55 58 52 52 17 16 1 ...
##  $ Hatch          : Factor w/ 2 levels "E","L": 1 2 2 1 2 ..
##  $ Pred           : Factor w/ 3 levels "C","L","NL": 3 1 ...
##  $ Res            : Factor w/ 2 levels "Hi","Lo": 1 1 1 1 ..
##  $ Age.DPO        : int  35 35 35 35 36 36 36 39 39 39 ...
##  $ Age.FromEmergence : int  1 1 1 1 2 2 2 5 5 5 ...
##  $ SVL.initial    : num  18 17.7 18.1 16.8 18.7 17.5 17.3 .
```

```
## $ Tail.initial     : num  5.4 1.1 5 6.4 6.3 4.4 1.3 1.5 2 .
## $ SVL.final        : num  17 18 17.8 17.1 19.3 17.8 17.9 ..
## $ Mass.final       : num  0.38 0.35 0.41 0.3 0.46 0.3 0.42
## $ Resorb.days      : int  3 3 3 3 3 4 2 2 2 3 ...
```

We can see that our dataset has 2502 observations of 14 different variables, some of which are integers, some are factors, and some are numerical. Things to notice:

1. Several variables are listed twice but are coded in different ways. For example, there is a column titled **Tank** and one titled **Tank.Unique**. As stated earlier, there are 12 tanks in each of 8 blocks. The variable **Tank** lists what number a tank is (1–12) in a given block, whereas **Tank.Unique** gives each tank a unique number out of the entire 96.

2. Similarly, we have the columns **Age.DPO** and **Age.FromEmergence**. The first one, **Age.DPO** is the age at emergence from the water in terms of days post-oviposition (DPO), whereas **Age.FromEmergence** counts the day the first animal crawled out of the water as Day 1, and so the age of animals is recorded in terms of days relative to when emergence began. Sometimes it can be useful to view the same data in two different manners.

3. We have three categorical predictors (factors): Hatching age, Predator treatment, and Resource level. Each factor has several levels, or categories, which you can see in the **str()** output.

4. We have several response variables (e.g., *short-vent length* (SVL) or Mass) that were measured at the initial point when froglets left the water, at the end of metamorphosis when the tail was fully resorbed, or both.

3.3.2 Data exploration and visualization

Begin by plotting the data to check for errors. The default plot function will create a simple graphic based on the type of data you provide it. In order to access a named variable within a data frame, we use the $ operator, as in *data_frame$variable*. The data frame always goes first, then the "$" then the name of the column you are interested in. For example, you might type in the following to start looking at your data.

```
plot(RxP$SVL.initial)
plot(RxP$Tail.initial)
plot(RxP$Mass.final)
plot(RxP$Pred)
```

What did that tell us?

1. SVL generally varied between 14 and 24 mm, but there is one animal that was much smaller.
2. Tail length at emergence varied from about 0 to 15 mm in most individuals, but there were three froglets with tails longer than 15 mm.
3. Mass varied from about 0.2 g to over 1 g.
4. The number of individuals surviving to metamorphosis in the three Predator treatments varied considerably, from over 1200 froglets in the Control group to approximately 500 in the Nonlethal group.

Notice that the style of plot changed depending on if we plotted a continuous variable or a factor (see Figure 3.1). For the continuous variable, the default is to plot the data in order, from the first row to the last. In the case of a factor, the default is to plot the number of observations in each group.

3.3.3 Further data exploration and identifying mistakes

Plotting data by itself can be useful, for example, if we want to check for outliers or find typos (which might make a numeric variable plot as a factor, for example). However, it is often more useful to plot response data against an explanatory variable. For example, maybe we want to know how the final mass of metamorphs varies across predator treatment. Here, we will use

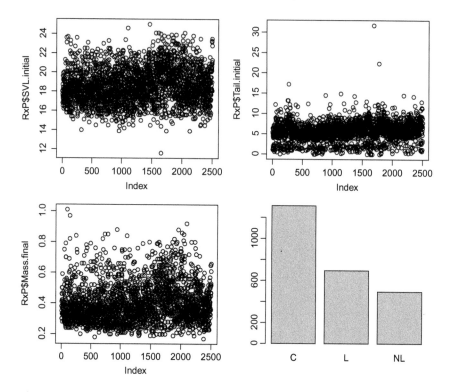

Figure 3.1 Plots of various single response variables.

the "~" to separate our response, or dependent, variable from a predictor, or independent, variable. Let's examine the relationship of **Mass.final** and **Predator** treatment by plotting "*Mass.final~Pred*".

```
plot(Mass.final~Pred, data=RxP)
```

Here are some important things to take note of. By providing a categorical variable as our predictor, R automatically knew to make a box-and-whisker plot (aka "boxplot"). There are not that many instances when R will think for you, but this is one.

Looking at Figure 3.2, there are several things to know about how R draws a boxplot. First, the top and bottom of each box represents the *interquartile range*, i.e., the middle 50% of your data. Thus, 25% of metamorphs in each predator treatment are larger than the top of their

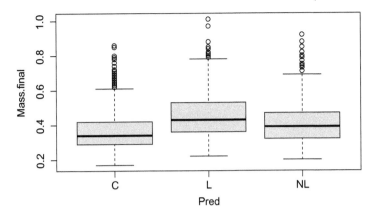

Figure 3.2 Relationship between predator treatment and mass at the end of *A. callidryas* metamorphosis.

respective box, and 25% are smaller than the bottom. Second, the heavy dark line in the middle of the box is the *median*, not the mean as many folks might initially think. Third, the extremes of the "whiskers" are either 1) the max or min value of the data or 2) 1.5 times the interquartile range (IQR). In the event of option #2, R will plot all the points that fall beyond the 1.5 times the IQR. So, what does that mean in practice? If you look at Figure 3.2, you can see that the bottom whiskers are all just that, a whisker. That means they have been plotted to the smallest value in the dataset and that that value falls within 1.5 times the IQR. The upper whiskers have a whole lot of points above them, meaning that the whiskers extend to the 1.5 times the IQR mark, and the points plotted above the whisker fall outside that range. Lastly, notice that we have introduced a new syntax. We can use the "~" to denote a relationship between two vectors, usually thought of as "response~predictor." This structure will be used later for defining statistical models and can be expanded to incorporate multiple predictors, e.g., "response~predictor1 + predictor2 + etc.".

What happens if we plot two continuous variables against one another, instead of a continuous response vs a categorical predictor? Maybe we want to see if there is a relationship between mass at the end of metamorphosis and SVL at the end of metamorphosis (Figure 3.3). Since we have provided two continuous variables, R will know to automatically make a scatterplot.

```
plot(Mass.final~SVL.final, data=RxP)
```

Figure 3.3 tells us several things.

- There appear to be several outliers. These are either individuals that have a very small SVL but a large mass, or vice-versa. These almost certainly represent mistakes that were made during data entry, since they are biologically unrealistic and maybe even are impossible, and they should be removed.
- The relationship between SVL and mass is not linear. It curves upward, which indicates that longer frogs (greater **SVL.final** values) seem to have disproportionately larger masses. This is to be expected in many length-to-mass relationships in nature, and perhaps plotting the data on logarithmic axes would make this relationship linear.

First, let's see if plotting the figure on log-log axes makes the length-to-mass relationship linear. There are two easy ways to accomplish this. Either

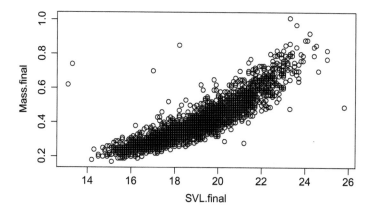

Figure 3.3 Relationship between metamorph SVL and mass at the end of *A. callidryas* metamorphosis.

bit of the following code would work. The first one will take the log of each variable and plot them against one another. Thus, the values on the axes will be in terms of the logarithm of either SVL or mass. The second example will plot the figure with normal numeric axes, but the scale of the axes will change at a log rate (Figure 3.4). Note that by denoting "*log*='*x*'" or "*log*='*y*'" you can just log transform one axis.

```
plot(log(Mass.final)~log(SVL.final), data=RxP)
plot(Mass.final~SVL.final, data=RxP, log="xy")
```

The two plots in Figure 3.4 are identical except in their axes. In the figure on the left, the axis is linear in its increments, but the values are log-transformed. In the figure on the right, the units correspond to the raw values that were measured, but they increase non-linearly. Comparing these figures with Figure 3.3, it indeed appears that log-transformation made the data more linear. If you're wondering why some numbers are missing from the axes, it is because R won't plot numbers on top of other numbers by default, and so makes some decisions about what to include vs what to exclude when spacing gets too tight.

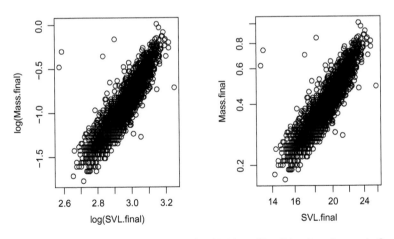

Figure 3.4 Relationship between log(SVL) and log(Mass) at the end of *A. callidryas* metamorphosis.

3.3.4 Fixing mistakes in the data

How do we go about fixing mistakes that we have found in the data file? One option would be to go back to your original .csv file or use Microsoft Excel (or whatever program you use to organize and input your data) and change it there. You should certainly do that! But this dataset has over 2500 individuals and trying to figure out which individuals are typos could be very tedious.

First, we can see in Figure 3.3 that there are four individuals that are recorded as having an SVL below 19 mm, and a Mass above 0.6 g. Thus, one way to identify where in the data file those mistakes are located is to search for any froglets with a Mass above 0.6 and an SVL below 19. We can do this using the []s, which is called indexing and allows us to subset the data based on some criterion or criteria.

3.3.5 A note about indexing

Using square brackets ([]s) to navigate a data frame (or other type of object) is one of the single most important things you can learn to do in R. If you think about a data frame as a two-dimensional object with rows and columns, every item (i.e., cell) in the data frame has a location in terms of its row and column. []s allow you to navigate the data in a simple and elegant manner. You did a tiny bit of this in Chapter 1, but here it is explained more fully. Square brackets allow you to find the location of any object in terms of [**row, column**]. Thus, if you want the object in the 5th row and the 3rd column in our data frame RxP, you would type the following:

```
RxP[5,3]
```

```
## [1]  10
```

You can also use []s to identify a range of things, like the first four rows and the first three columns.

```
RxP[1:4,1:3]
```

```
##    Ind Block Tank
## 1    1    5    7
## 2    2    5    4
## 3    3    5    4
## 4    4    5    7
```

Most importantly, you can enter in **logical statements** within the []s. For example, if you want to return all of the rows that are from the Low resource treatment, you can type the following code. Note that the column part of the [row, column] notation is left blank, which indicates to return every column in the data frame. Also, note the code to define what you want to return. The following code states, "return every row of RxP where the Resource column is equal to Lo." I won't display it here, since it is A LOT of output, but you should type it in to your console and see what happens.

```
RxP[RxP$Res=="Lo",]
```

┤ Box 3.4 - Logical statements ├

In R, like pretty much any programming language, there exist a series of simple coding phrases, or operators, which allow you to define logical statements. These operators can be extremely powerful when used correctly. The R logical operators are the following:

Symbol	Meaning
<	is less than
<=	is less than or equal to
>	is greater than
>=	is greater than or equal to
==	is equal to
!=	is not equal to
|	or
&	and

Using logical operators, you can do all kinds of powerful things, but the most common is probably to subset your data. Imagine we wanted to create separate objects that contained the Lo and Hi resource individuals and store those as new objects. To do so, we could type the following:

```
RxP.Lo<-RxP[RxP$Res=="Lo",]
RxP.Hi<-RxP[RxP$Res=="Hi",]
```

Let's use indexing to find our errors and remove them from the data frame. For example, if we want to identify every froglet that has a final SVL smaller than 19, we can type the following:

```
RxP[RxP$SVL.final<19,]
```

Now, some folks find the square brackets a little difficult to wrap their heads around, so, I'm going to switch to using a function called **filter()** from the *dplyr* package. In the coming pages and chapters, you will read a lot about *dplyr* and other assorted functions that make up what is known as the *tidyverse*. *dplyr* is an amazing package and we will begin to explore some of its possibilities later in this chapter. The main advantage of **filter()** is that it allows you to write the operation in a slightly more intuitive language than the [rows,columns] notation of square brackets. To execute the same operation as previously shown with **filter()**, type the following (and don't forget to load *dplyr*!):

```
library(dplyr)
filter(RxP, SVL.final<19)
```

This tells R to give us every row of RxP where values in the column SVL.final are smaller than 19. If you executed this line, you would see this returns a lot of individuals! The same logic applies if we want to subset the data based on multiple criteria. We can string criteria together using the "&" logical command. For example, we want to find every froglet with an SVL smaller than 19 mm and a mass greater than 0.6 g.

```
##
## Attaching package: 'dplyr'
```

```
## The following objects are masked from 'package:stats':
##
##      filter, lag

## The following objects are masked from 'package:base':
##
##      intersect, setdiff, setequal, union
```

```
filter(RxP, SVL.final<19 & Mass.final>0.6)
```

```
##      Ind Block Tank Tank.Unique Hatch Pred Res Age.DPO
## 1   734     2    7           19     E   NL  Hi      51
## 2  1024     8    7           91     E    L  Lo      48
## 3  1078     6   11           71     E    C  Lo      54
## 4  1284     1    2            2     E    C  Hi      62
##    Age.FromEmergence SVL.initial Tail.initial SVL.final
## 1                 17        18.0          0.6      18.2
## 2                 14        22.2          8.3      13.3
## 3                 20        16.2          1.5      17.0
## 4                 28        22.1          9.1      13.1
##    Mass.final Resorb.days
## 1        0.85           1
## 2        0.74           6
## 3        0.70           3
## 4        0.62           6
```

This command returned just four individuals, which correspond to the four points plotted in the upper left corner of the plots we saw earlier. Looking at Figure 3.3, there appears to be one other obvious point that has a suspiciously low mass despite having the largest SVL. We can see it has an SVL larger than 24 mm and a final mass smaller than 0.6 g, so let's use those as our filters.

```
filter(RxP, SVL.final>24 & Mass.final<0.6)
```

```
##      Ind Block Tank Tank.Unique Hatch Pred Res Age.DPO
```

```
## 1 1127      1    3              3    L    C   Hi        58
##    Age.FromEmergence SVL.initial Tail.initial SVL.final
## 1                 24        20.3          4.7      25.8
##    Mass.final Resorb.days
## 1       0.49           3
```

Now we have identified five individuals that look like they are the result of bad data entry. This is likely to happen in any large experiment, so it is important to clean up the data before analyzing it.

At this point we are ready to remove these individuals from our data frame forever, which we can do with **filter()**. Our problem individuals are 734, 1024, 1078, 1127, 1284, which is indicated by the row numbers shown on the left side of the output as well as the "*Ind*" column, which is each individual's unique ID in the dataset. We want to get rid of these rows entirely, which we can do by using the logical command '!=', which means "not equal to." Lastly, let's create a new data frame called "*temp*" that excludes the data we don't want. I like to call things "temp" when I know it won't be around for long and I will rename it once I get it to the final state I want it in.

```
#Remove the offending data
temp<-filter(RxP, Ind!=734 &
             Ind!=1024 &
             Ind!=1078 &
             Ind!=1127 &
             Ind!=1284)
```

The previous command essentially subsets the RxP data frame and returns every column when the Individual is not equal to 734, 1024, 1078, 1127, or 1284, and then assigns that resulting data frame to the object temp. Notice that R does not provide any warning messages if you write over data. *It is very easy to accidentally write over your data!* That is one reason why we made a new data frame. We should confirm that it worked by plotting the data (see Figure 3.5).

```
plot(Mass.final~SVL.final, data=temp, log="xy")
```

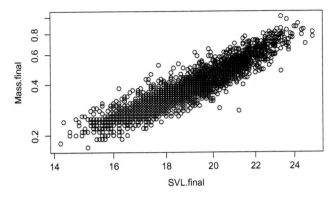

Figure 3.5 Relationship between log(SVL) and log(Mass) at the end of *A. callidryas* metamorphosis, with instances of obvious data entry mistakes removed.

⊢ Box 3.5 - Outliers? ⊢

Identifying outlier data is often difficult and subjective. What one researcher might see as an obvious outlier, another researcher may not. Such is life. One reason to remove outliers, defined here as infrequent and extreme values, is that they can have tremendous influence on your statistical analyses.

A good way to see if outliers have undue influence on your model is to run the model with and without the data in question and see if it changes the result. If removing the one or two data points does not affect anything, then you can probably feel comfortable leaving them in. But if those data points change the result, then you have to decide what to do. Perhaps, as in the example of the RxP dataset, you have a lot of data (here we have more than 2500 individuals). If one or two individuals change the results for the other 2500 froglets, then they are certainly exerting too much influence and our ability to estimate the ecological relationships of interest will be faulty.

By plotting various response variables against one another, we can further explore the data and search for more errors or potentially find genuine outliers. Some examples of this might include the following (Note that we are now working with our modified data frame **temp**):

```
plot(SVL.initial~Age.DPO, data=temp)
plot(Mass.final~Age.DPO, data=temp)
plot(Resorb.days~Tail.initial, data=temp)
plot(SVL.final~SVL.initial, data=temp)
```

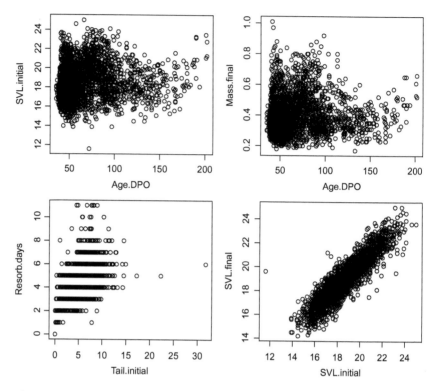

Figure 3.6 Various response variables plotted against one another in order to look for outlier data or obvious data entry mistakes.

Figure 3.6 tells us a lot of things. First, in the fourth panel there is one froglet that has a recorded initial SVL that is much smaller (less than 12 mm) than any other froglet. This seems to clearly be an error, especially given that the final SVL of that same individual is about 19 mm! Secondly, although nearly all froglets had tails between 0 and 15 mm, three had tails longer than this. These might be real data points but they 1) are not representative of the overall population and 2) have a lot of potential to skew our results since they are so extreme (which they would do if we were to leave them in the dataset).

First, let's identify the froglet with the incorrect very small initial SVL and remove it by re-assigning temp without the problem individual to the same object temp. This is essentially like writing over a file with a new version of itself. Since we know these data can be identified with a single logical statement, we can easily remove them by subsetting the data frame to everything that *is not the data we want to get rid of.* For example, if we want to get rid of the one data point that is smaller than 12 mm, we can just re-assign temp to be everything larger than 12 mm.

```
filter(temp, SVL.initial<12)
```

```
##      Ind Block Tank Tank.Unique Hatch Pred Res Age.DPO
## 1 1655     6    8          68     E    C  Hi      72
##    Age.FromEmergence SVL.initial Tail.initial SVL.final
## 1                38        11.6          5.3      19.6
##    Mass.final Resorb.days
## 1       0.43           5
```

```
temp<-filter(temp, SVL.initial>12)
```

The same logic applies to identifying and removing the three individuals with the very long tails.

```
filter(temp, Tail.initial>15)
```

```
##      Ind Block Tank Tank.Unique Hatch Pred Res Age.DPO
## 1  271     1    2           2     E    C  Hi      44
## 2 1689     2   10          22     E    C  Hi      76
## 3 1777     7    8          80     E    C  Lo      76
##    Age.FromEmergence SVL.initial Tail.initial SVL.final
## 1                10        18.8         17.2      18.9
## 2                42        23.3         31.7      23.7
## 3                42        15.8         22.3      15.6
##    Mass.final Resorb.days
## 1       0.36           5
## 2       0.64           6
## 3       0.23           5
```

```
temp<-filter(temp, Tail.initial<15)
```

At this point, it is important for you, the reader, to make sure the version of the data that you are working with is the same as that shown in the following example in **temp**. You should have 2493 individuals and 14 columns (observations). If you don't, you should go back through the code and figure out what you might have missed. Once you know you are good to go, the dataset is ready to analyze! For good measure, let's rename our dataset **RxP.clean** so that it is clear that we are using the cleaned-up data with the outliers removed. All of the examples moving forward will use **RxP.clean** so it is important that you have the same data as the book!

```
str(temp)
```

```
## 'data.frame':    2493 obs. of  14 variables:
## $ Ind              : int  1 2 3 4 5 6 7 8 9 10 ...
## $ Block            : int  5 5 5 5 5 5 5 2 2 1 ...
## $ Tank             : int  7 4 4 7 10 4 4 5 4 1 ...
## $ Tank.Unique      : int  55 52 52 55 58 52 52 17 16 1 ...
## $ Hatch            : Factor w/ 2 levels "E","L": 1 2 2 1 2 ..
## $ Pred             : Factor w/ 3 levels "C","L","NL": 3 1 1 .
## $ Res              : Factor w/ 2 levels "Hi","Lo": 1 1 1 1 ..
## $ Age.DPO          : int  35 35 35 35 36 36 36 39 39 39 ...
## $ Age.FromEmergence: int  1 1 1 1 2 2 2 5 5 5 ...
## $ SVL.initial      : num  18 17.7 18.1 16.8 18.7 17.5 17.3 ..
## $ Tail.initial     : num  5.4 1.1 5 6.4 6.3 4.4 1.3 1.5 2 ...
## $ SVL.final        : num  17 18 17.8 17.1 19.3 17.8 17.9 ...
## $ Mass.final       : num  0.38 0.35 0.41 0.3 0.46 0.3 0.42 ..
## $ Resorb.days      : int  3 3 3 3 3 4 2 2 2 3 ...
```

```
#Assign temp to be an object with a useful name
RxP.clean<-temp
#Remove temp
rm(temp)
```

3.4 SUMMARIZING AND MANIPULATING DATA

There are many tools in R to help you summarize your data efficiently. I think there are two functions worth knowing about at this stage: **aggregate()** from base R and **summarize()** from the *dplyr* package.

aggregate() is a very useful function that takes a column of raw data and summarizes it across one or more groups based on some chosen function (e.g., calculate the mean or the standard deviation, etc.). One nice think about using **aggregate()** is that you code the function of how you want your data summarized *in the exact same format* that you use for specifying plots or models (which we will cover in Chapter 5). The output of **aggregate()** is a data frame, which makes it easy to then use for plotting figures or other purposes. **aggregate()** takes three arguments: 1) the response and predictor variables, 2) the function you want to execute (with the "*FUN=*" argument), and 3) the data frame where the data can be found. For example, if you want to calculate the mean size of metamorphs at emergence across all combinations of Food and Resource treatments, you can type the following:

```
#This is great way to use aggregate
aggregate(SVL.initial~Pred*Res, FUN=mean, data=RxP.clean)
```

```
##    Pred Res SVL.initial
## 1    C  Hi    18.37231
## 2    L  Hi    19.64221
## 3   NL  Hi    19.39076
## 4    C  Lo    17.87029
## 5    L  Lo    19.14696
## 6   NL  Lo    18.41102
```

The package *dplyr* has many useful functions for data wrangling which we will cover in this book. By using some of these in combination, you can easily summarize your data. The utility of this may not be completely evident now, but it will be later, I promise (particularly in Chapter 9). The downside of *dplyr* is that, much like *ggplot2*, it has its own lexicon that is fairly distinct from the rest of R, meaning that you have to learn a whole different set of commands. So be it. It's pretty great once you learn the coding.

By using a few choice functions, such as **group_by()** and **summarize()**, we can easily say we want to calculate the means of whatever variable (or variables) we are interested in. Note of course that you can use different functions besides **mean()**, which is what we will start with here. The other important thing to note in the following code is the use of the **pipe** command, '**%>%**', which designates the output of one line to be the input of the next.

Box 3.6 - The pipe command: %>%

The pipe is a function in and of itself and it is really important to understand. The pipe allows you to string together series of functions one after another after another. Thus, a single line of code is completed and produces some output and the next line of code automatically inherits that output as its input. We will really see the utility of the pipe in Chapter 9 on data wrangling, but hopefully you will grasp how handy it is in the examples in this chapter.

The following code does several things all at once, so let's walk through it up front. The first line does two things. First, it defines what data frame we are going to be using (*RxP.clean*) and then "pipes" it to the next line. It also states that we are going to store the output of the string of commands to an object called "*RP.means*." The second line uses the function **group_by()** to say we are going to be grouping the data by the resource and predator treatments, which are piped to the next line. Thus, the output of the second line is just the same data frame we started with, but now the data frame knows we want to think of two specific grouping variables (and all combinations of them) in any subsequent commands. Lastly, the third line says that we will calculate a new column and we will call it *SVL.mean*, and it will be calculated as the mean (which is of course a function) of the data in the column called *SVL.initial*. Since the **summarize()** function has inherited the grouped data frame from the previous line, it knows to find the data in the *RxP.clean* data frame and to calculate means of all the combinations of our Res and Pred groups. Don't forget to load the *dplyr* package if you haven't used it yet!

```
library(dplyr)
```

```
RP.means<-RxP.clean %>%  #First, identify the data frame
      group_by(Res, Pred) %>%  #Next, establish grouping variables
      summarize(SVL.mean = mean(SVL.initial)) #Last, calc. means
RP.means
```

```
## # A tibble: 6 x 3
## # Groups:   Res [2]
##    Res   Pred  SVL.mean
##    <fct> <fct>    <dbl>
## 1 Hi    C         18.4
## 2 Hi    L         19.6
## 3 Hi    NL        19.4
## 4 Lo    C         17.9
## 5 Lo    L         19.1
## 6 Lo    NL        18.4
```

A few other things to note. The object that was created by **summarize()** is called a "tibble," as in the pronounced version of the abbreviation "tbl." It is just like a data frame, but instead shows you the type of data in each column and only shows you the top 10 rows (if your data frame has more than 10). You can turn it in to a standard data frame by using the function **as.data.frame()** to redefine it. This is not only useful, but necessary, in some situations.

```
#Here is the structure of a tibble
#Note how it is different from a data frame
str(RP.means)
```

```
## tibble [6 x 3] (S3: grouped_df/tbl_df/tbl/data.frame)
##  $ Res     : Factor w/ 2 levels "Hi","Lo": 1 1 1 2 2 2
##  $ Pred    : Factor w/ 3 levels "C","L","NL": 1 2 3 1 2 3
##  $ SVL.mean: num [1:6] 18.4 19.6 19.4 17.9 19.1 ...
##  - attr(*, "groups")= tibble [2 x 2] (S3: tbl_df/tbl/data.fram
##   ..$ Res   : Factor w/ 2 levels "Hi","Lo": 1 2
##   ..$ .rows: list<int> [1:2]
##   .. ..$ : int [1:3] 1 2 3
```

```
##    .. ..$ : int [1:3] 4 5 6
##    .. ..@ ptype: int(0)
##    ..- attr(*, ".drop")= logi TRUE
```

```
#Convert the tibble output to a data frame
RP.means<-as.data.frame(RP.means)
str(RP.means)
```

```
## 'data.frame':    6 obs. of  3 variables:
## $ Res     : Factor w/ 2 levels "Hi","Lo": 1 1 1 2 2 2
## $ Pred    : Factor w/ 3 levels "C","L","NL": 1 2 3 1 2 3
## $ SVL.mean: num  18.4 19.6 19.4 17.9 19.1 ...
```

3.4.1 Reordering your data

In Chapter 4, we are going to create a bar graph of mean metamorph size with standard error bars. Before we do that however, you might have noticed by now that R orders the levels of a factor alphabetically. Thus, for our predator treatment we can see that "C" comes before "L," which comes before "NL." For plotting purposes, it would probably make more sense to have them ordered Control, Non-lethal, Lethal, or vice-versa. That at least makes sense to me! To see the current ordering of the levels of any categorical variable, use the function **levels()**.

```
#What is the current order of the factor levels?
levels(RxP.clean$Pred)
```

```
## [1] "C"  "L"  "NL"
```

As with most things in R, there are multiple ways to reorder the factor levels. Here are two examples. In both examples, note that we have to assign the releveled object to replace itself in the data frame.

1. One way to change the order would be use the function **factor()** to recode all the levels of the factor at once. **factor()** is a generic function which could be used, for example, to turn a vector of text or of 0's and 1's into factor levels. Here, we are essentially

writing over the current factor with a new factor that is the same but has a different order for the factor levels. R knows to look at the values in the column and see how many possible levels there are (for example, if you provided just two levels in the code, R would give you an error because it can see that there are three different categories present).

```
#This is one way to reorder a factor
RxP.clean$Pred<-factor(RxP.clean$Pred, levels=c("C","NL","L"))
```

Note that we did not change the data at all, just the way the levels of the factor are considered.

2. A second way is to use the function **mutate()**, which is how you create new variables in the world of *dplyr*.

```
#This is another way to reorder a factor, using dplyr
RxP.clean<-RxP.clean %>%
  mutate(Pred = factor(Pred, levels=c("C","NL","L")))
```

Why would you choose to use one method over another? Both techniques use the same function, **factor()**, to do the actual reordering of the levels. Thus, it really just boils down to personal preference. If you get really used to the *dplyr* style of coding, then you might feel most comfortable using **mutate()** but otherwise the first option is a little shorter to type.

3.4.2 Calculating Means and Standard Errors

Now that our Pred factor is in the order we want, let's use *dplyr* to make a new data frame which contains the means and standard errors of the data. We can just modify our code from earlier where we made "*RP.means*." There is no built-in function to calculate the standard error in R (there are multiple definitions of "standard error" and the R gurus could never agree on a single definition of what exactly a standard error is), so we will have to calculate it ourselves. For this exercise, we will use the most common and widely accepted definition: standard deviation/square root of N.

The following code utilizes three *dplyr* functions: **group_by()**, **summarize()**, and **mutate()**. It is important to remember that since each line inherits the output of the previous line via the pipe command, by the time we get to **mutate()** the object to be manipulated is the summarized data from the previous line. Thus, the object being input into **mutate()** is fundamentally different from the starting object ("*RxP.clean*") and did not exist before running **summarize()**. Note that if you were to change the order of *Res* and *Pred* in the **group_by()** function, it would change the order of how the data are output. Go ahead and try it and see what happens.

```
RP.means<-RxP.clean %>% #Define what dataset we are using
      group_by(Res, Pred) %>% #Set the groups
      summarize(SVL.mean = mean(SVL.initial), #Calculate the mean
            SVL.sd = sd(SVL.initial), #SD
            SVL.n = length(SVL.initial)) %>% #and N of SVL
      mutate(SVL.se = SVL.sd/sqrt(SVL.n)) #Calculate the SE
RP.means
```

```
## # A tibble: 6 x 6
## # Groups:   Res [2]
##    Res    Pred   SVL.mean SVL.sd SVL.n SVL.se
##    <fct>  <fct>     <dbl>  <dbl> <int>  <dbl>
## 1 Hi     C          18.4   1.80   679 0.0692
## 2 Hi     NL         19.4   1.72   249 0.109
## 3 Hi     L          19.6   1.62   398 0.0812
## 4 Lo     C          17.9   1.74   626 0.0697
## 5 Lo     NL         18.4   1.68   245 0.107
## 6 Lo     L          19.1   1.74   296 0.101
```

Okay, so now we have a nice, tidy summarized version of our output. We can see that it has the mean, standard deviation, sample size (n) and standard error of the initial SVL of froglets when they crawled out of the water. We will use this object ("*RP.means*") in the next chapter for making a bar graph.

Box 3.7 - Take-home message

- Whenever you import data into R, always spend some time exploring it to check for errors and to make sure that everything imported correctly. Looking at the structure of your data frame with **str()** is one of the fastest ways to see if you had typos in your original data file.
- To explore your data, you should also make scatterplots, box-and-whisker plots, histograms, density plots, etc. It is crucial to look at your data from many different perspectives to see if anything looks out of the ordinary.
- Whether you choose to use square brackets (**[]**) or the **filter()** function, learning how to use logical operators is crucial to being able to efficiently navigate and subset your data as needed.
- **dplyr** is a fantastic package for summarizing your data quickly and efficiently.

3.5 ASSIGNMENT!

Here are some things to do on your own. Check the Github page for example answers and code at https://github.com/jtouchon/Applied-Statistics-with-R.

1. Use **filter()** to subset the data to just those froglets that emerged during or after the first 30 days of the period of metamorphosis.
2. Next, create box-and-whisker plots of SVL.initial or Mass.final for each group. What does this reveal about animals that emerged earlier or later in the window of metamorphosis?

Introduction to Plotting

4 Introduction to Plotting

The purpose of this chapter is to introduce you to making excellent, professional quality figures from your data. We are covering this now because we will be doing more and more plotting in the coming chapters.

4.1 PRINCIPLES OF EFFECTIVE FIGURE MAKING

Before we get into the meat of how to most efficiently plot your data, it is useful to take a moment to talk about what makes a good figure. What is it that makes a high-quality figure, one that is suitable for publication or use in a presentation? I would argue that judicious use of color, large clear text and labels, and efficient usage of plot space are three hallmarks of a good figure. It is often useful to create multiple panels to show different aspects of your data. These fundamentals are the same whether you are making your figures in R or not or using base graphics or **ggplot2**. If you are interested in a deeper dive into this topic, please check out the excellent book *Fundamentals of Data Visualization* by Claus O. Wilke, professor at The University of Texas (UT) Austin. He has written a particularly elegant and useful package called **cowplot** that provides not only a nice and easy to apply theme for **ggplot2** to make

Applied Statistics with R: A Practical Guide for the Life Sciences. Justin C. Touchon, Oxford University Press (2021). © Justin C. Touchon. DOI: 10.1093/oso/9780198869979.003.0004

your figures look betterm, but it also contains many other functions which are useful.

There are two main ways to make figures in R: base graphics (i.e., those that are built into the base version of R you downloaded from the Comprehensive R Archive Network (https://cran.r-project.org/) (CRAN) and using the package *ggplot2*. There are functions in other packages (e.g., the **scatterplot()** function in the *car* package, or the **barplot2()** function in the *gplots* package), but these all utilize the basic coding of base graphics. While *ggplot2* and base graphics use different coding styles, the fundamentals that make an effective graphic remain the same. Both types of coding allow you to build your graphics piece by piece and really give you control over every aspect of the figure.

Let's talk about some basic nuts and bolts that are useful to know.

4.1.1 Defining colors

As with so many things in R, there are multiple ways to define what colors you want to use. Each system has its pros and cons.

4.1.1.1 Numerical colors

If you are just making a simple plot to explore your data, the easiest way to define color is by just using one of the nine basic numbered colors (e.g., "*col=1*"). These are as follows: 0=white, 1=black, 2=red, 3=green, 4=blue, 5=light blue, 6=fuschia, 7=yellow, 8=grey. If you name a color higher than 8, the palette wraps back around to 1. Thus, 9 is the same as 1 (black), 10 is the same as 2 (red), and so on.

4.1.1.2 Named colors

R has 657 named colors stored and ready to use. The full list of these can be viewed by typing **colors()** (or if you prefer, **colours()**) at the command prompt. Using named colors is nice because you probably have an intuitive sense of what "slate grey" is going to look like, whereas a color defined by its red, green, and blue (RGB) values is less obvious. If you need to know what the exact RGB values of a named color are, simply use the function

col2rgb() and place the name of the color in quotes in the parentheses. For example, "*col2rgb('yellow')*" would tell you the RGB values for yellow in R.

4.1.1.3 RGB colors

R allows you to define colors based on their red, green, and blue (RGB) values. This is particularly useful if you have a color scheme you want to match, perhaps for a presentation in MS PowerPoint or Apple Keynote, and thus you can specify the exact RGB color you are looking for (which might have come from that other program). Colors are defined using the **rgb()** function, which requires 3 arguments (not surprisingly they are the red, green, and blue values) as well as other optional arguments. The default is that all values have a minimum of 0 and a maximum of 1, but you can set the max value to be 255 if you wish (which is likely how colors are defined in another program of your choosing). Lastly, you can also make an RGB color translucent by adjusting the alpha level (note that alpha is always on the same scale as the RGB values). For example, to set a color that was pure red but was 30% translucent (or conversely, 70% opaque), you would type the following.

```
#Sometimes it is nice to define colors using the RGB system
col=rgb(red=255, green=0, blue=0, maxColorValue=255, alpha=180)
```

4.1.1.4 Hex colors

Technically, R actually stores color in a system called hexadecimal, which defines each color (red, green, or blue) in terms of two values that range from 0–9, then from A–F, giving a total of 16 different values for each character. Thus, since each color is defined in terms of two number/letter combos, each color has 256 different possible values (the same as we see with RGB!!). Although the code for hexadecimal may not be intuitive, it is easy to look up online exactly what the code is for any color you want to use (just do an internet search for something like "RGB to Hex color"). The main advantage of the Hex system over RGB is that it is very compact to specify whatever color you want. In R, the hex color code goes in quotes and is preceded by a #. An additional two characters can be added to define a degree of translucency (aka, the alpha level). For example, the translucent red color defined previously would be as follows:

```
#This is how you define a color using the hex system
col="#FF0000B4"
```

4.2 DATA EXPLORATION USING *ggplot2*

In the last chapter, you learned a little bit about how to make fairly simple figures using base graphics, i.e., the graphics functions that are built into the version of R you downloaded from CRAN. One problem with base graphics is that the figures produced are relatively utilitarian and ugly (at least in many people's view). There is a whole universe of functions and arguments you can use to make them look better, but in their basic version they are kind of boring and ugly. The relatively recently designed package *ggplot2* makes it very easy to make nice looking figures. However, the syntax for coding in *ggplot2* is a little different than base graphics. There are two workhorse functions in *ggplot2*: the first is **qplot()** (which stands for "quick plot") and the other is **ggplot()**. We will cover **ggplot()** at a later date. For now, let's explore **qplot()**.

Box 4.1 - A Quick Note About the Pros of *ggplot2*

I taught myself R back in the dark ages of the 00s, and so learned how to make nice looking plots from scratch. Thus, when **ggplot2** came along, I was hesitant to jump on board and learn a lot of new code. Let me assure you dear reader that it is worth your efforts. Here is a short list of what I see as the benefits of using **ggplot2** and its functions.

- Makes nice looking graphics much easier than base graphics
- Allows easy data exploration via faceting
- You can assign your plots to an object, and the object contains all the data and code necessary to make the plot (i.e., it is more than just an image)
- Dovetails with other **tidyverse** packages like **dplyr** to allow you to seamlessly wrangle your data and plot it at once

I will introduce **ggplot2** bit by bit in each chapter and with much more depth covered in Chapter 9. That said, we will only cover the essentials of **ggplot2** in this book, so you are strongly encouraged to purchase the excellent book *ggplot2: Elegant Graphics for Data Analysis* by Hadley Wickham, Danielle Navarro, and Thomas Lin Pedersen.

4.2.1 Boxplots

Earlier, you made a boxplot using base graphics. The syntax for this was conveniently the same that we will use to define statistical models ("response~predictor"). **ggplot2** does things differently. Instead, you explicitly specify what variable you want on the x or y axes, and you specify the type of plot you want using the *geom* argument, which is short for *geometric object*. Don't forget that in order to use the **qplot()** function you first need to load the **ggplot2** library. Also remember that you have to do this each time you restart R. In the following code I am using the number sign (#) to annotate the code. When you come back to your analyses in a week, a month, or a year, you need to have notes to remind yourself of what you were doing. Always leave notes in your script file for your future self.

```
#Load the necessary package
library(ggplot2)
#Make a boxplot using qplot
qplot(data=RxP.clean,
      y=Mass.final,
      x=Pred,
      geom="boxplot")
```

Even in the most basic form, the figure made with **ggplot2** (Figure 4.1) is nicer looking (to many people), but this is not why **ggplot2** is so useful. Where **ggplot2** really shines is in its ability to add colors and to plot data across many different variables at once, as well as to easily take the same type of data and plot it in different ways within a single coding framework. For example, if you want to add colors to your figures, you can use either the "*col*" or "*fill*" arguments. In the case of a boxplot, "*col*" will change the color of the outline of the boxes whereas "*fill*" changes the color inside each one. The effects of "*col*" or "*fill*" will differ based on the particular type of plot you are making. One thing that is really cool about plotting with **ggplot2** is that we define the colors as one of the variables in our dataset. R will be able to look at our data frame and know how many categories we have, and therefore how many colors to plot, and will even add a legend for us. So handy!

Here, I've also loaded in the package **cowplot** that has some handy features, one of which is the function **plot_grid()** which allows you to plot

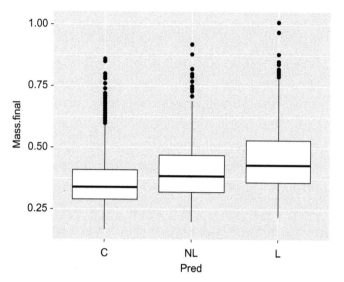

Figure 4.1 A boxplot showing the final mass of *A. callidryas* metamorphs plotted against predator treatment. This plot was made with **_ggplot2_**. For comparison, look at Figure 3.2 which are the same data plotted with base graphics.

multiple figures in a single window when using **_ggplot2_**. Just FYI, **_cowplot_** has nothing to do with cows *per se*, it was written by Claus O. Wilke, whose initials spell "cow," and who is a professor at the University of Texas Austin, whose mascot also happens to be a cow. There are other ways to plot multiple figures in a single plot window using base graphics, but we won't bother to discuss them here. Lastly, I've assigned each plot to be an object (another one of the cool features of **_ggplot2_**), to make it easier to plot multiple figures together.

```
library(cowplot)
#Change the outline color of the boxes and whiskers
a<-qplot(data=RxP.clean,
         y=Mass.final,
         x=Pred,
         geom="boxplot",
         col=Pred)
#Change the color that fills each of the boxes
b<-qplot(data=RxP.clean,
```

```
        y=Mass.final,
        x=Pred,
        geom="boxplot",
        fill=Pred)
plot_grid(a,b, ncol=2)#specify to put the plots in two columns
```

4.2.2 Faceting

Okay, so you can add colors really easily (Figure 4.2). What if we want to plot our mass data across multiple predictors, to more fully explore how the data fall out across different groups? We can easily do this by adding in the "*facets*" argument. Faceting allows you to easily split your plot window in an intuitive manner. The syntax for facets is "*rows~columns*." In other words, you specify the variable you want in each row and what you want in each column. If you only want to facet in one direction, you need to place a " . " on the other side of the "~".

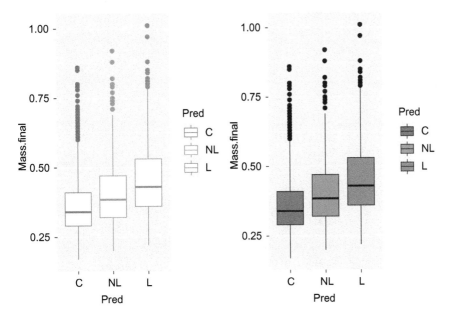

Figure 4.2 It's really easy to add nice looking colors with **ggplot2**!

```
#Plot the mass data by all three categorical
#predictors by using facets
qplot(data=RxP.clean,
      y=Mass.final,
      x=Pred,
      geom="boxplot",
      fill=Pred,
      facets=Hatch~Res)
```

4.2.3 Violin plots

Boxplots are great for seeing some basic info about the spread of data you have, but they can still be somewhat misleading because they don't show how the data are distributed within the box. One alternative is known as a violin plot. The coding is the same as for a boxplot, but the "*geom*" is now defined as "*violin*" instead of "*boxplot*." The following code also exemplifies one other great thing about **ggplot2**. By defining the "fill" argument as a variable (not already defined as our x-axis), we can automatically plot the various categories within the treatment (Figure 4.4). So convenient!

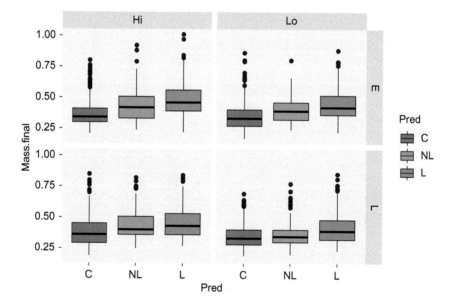

Figure 4.3 Whoa!! It was so easy to break our data apart into subplots with **qplot()**!

```
#Plot the mass data as a violin plot
qplot(data=RxP.clean,
      y=Mass.final,
      x=Pred,
      geom="violin",
      fill=Res)
```

4.2.4 Histograms and density plots

By default, if you plot just a single continuous variable using **qplot()**, R will plot a histogram (Figure 4.5, left). Histograms are extremely useful for seeing the distribution of your data. Do they look normally distributed? Are they skewed to one side or the other? There is technically no need to specify a "*geom*" here, but I think it is a good idea to be clear in your code, so I would recommend it.

```
#Make a basic histogram
a<-qplot(data=RxP.clean,
         x=Mass.final,
         geom="histogram")
```

The same principle of using the color or fill arguments as a way to view your data apply to histograms, but with one caveat. If you add a fill or color argument to a histogram in **qplot()**, R will make a *stacked* histogram (Figure 4.5, middle). It can be more useful to see the data distributions overlayed on one another. This is best achieved with a *density plot*, which is similar to a histogram but instead plots a smoothed line that shows the shape of the data (Figure 4.5, right). Note that you should use the "*col*" argument in the density plot instead of the "*fill*" argument. What happens if you do not?

```
#Make a stacked histogram
b<-qplot(data=RxP.clean,
         x=Mass.final,
         geom="histogram",
         fill=Pred)
#Make overlayed density plots
c<-qplot(data=RxP.clean,
         x=Mass.final,
         geom="density",
```

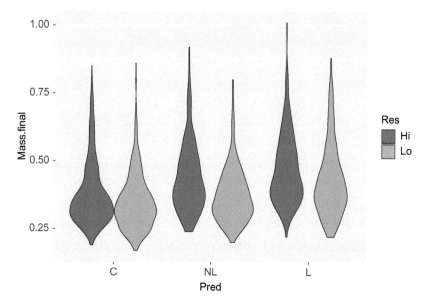

Figure 4.4 Violin plots of final mass of metamorphs across resource and predator treatments. The violin plots make it easy to see where the preponderance of data are. For example, we can see that in the Nonlethal predator treatment, the Low resource treatment has a slightly downshifted distribution compared to the High resource treatment.

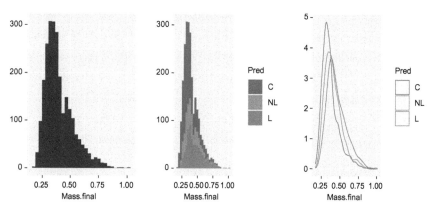

Figure 4.5 Three ways to view the distributions of the data, the final mass of *A. callidryas* metamorphs, made with **qplot()**. The plot on the left is a standard histogram. The middle plot is a stacked histogram, showing which data contribute to the overall shape of the histogram. The plot on the right is a density plot. You could of course also use facets to split the three treatments into separate panels.

```
        col=Pred)
#Plot all three figures together
plot_grid(a,b,c,ncol=3)
```

4.2.5 Scatterplots

The same principle works for continuous response variables. Previously, we defined our x-axis as a categorical variable, but if we instead use a continuous variable R will plot a scatterplot. We can still use facets or colors to visualize the variation in our data, which is extremely useful. For example, in the following code I've filled the points based on the resource treatment, and faceted the data based on the predator treatment. Imagine the possibilities (Figure 4.6)!

```
#Make a series of scatter plots
qplot(data=RxP.clean,
      x=log(SVL.final),
      y=log(Mass.final),
      col=Res,
      facets=.~Pred)
```

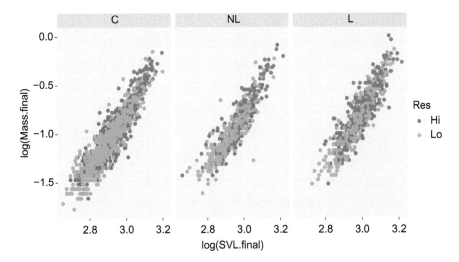

Figure 4.6 The log of the final mass of metamorphs plotted against the log of their SVLs, plotted with **qplot()**. The data are faceted by predator treatment and points are colored by resource treatment.

Note that when we make a plot like this, many of the points end up on top of one another, making it difficult to see all the data. We can add an argument to set the *alpha* level, or the degree of translucency of the points to alleviate this issue. This is also particularly useful with density plots. For example, we can remake the density plots from above, but this time we will fill them instead of color them and set the alpha to be 0.5 (Figure 4.7).

```
qplot(data=RxP.clean,
      x=Mass.final,
      geom="density",
      fill=Pred,
      alpha=0.5)
```

Note that setting the alpha level in **qplot()** makes the alpha level 50% transparent, no matter what value you enter. I will show you how to set it to whatever you want later.

In addition to visualizing our data by setting the fill or color to one of our variables, we can also change the shape of the points based on a variable in our data frame with the "*shape=*" argument (Figure 4.8).

```
#Make a series of scatter plots
qplot(data=RxP.clean,
      x=log(SVL.final),
```

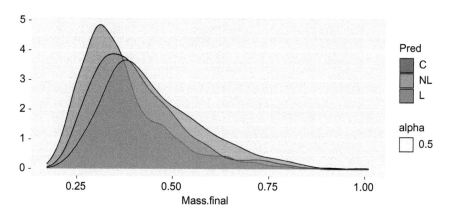

Figure 4.7 Here is another great way to visualize a distribution of your data, all with **qplot()**. Here we have made a density plot, but now we have made the fill color 50% transparent. Does that look great?

```
        y=log(Mass.final),
        col=Hatch,
        alpha=0.5,
        shape=Res,
        facets=.~Pred,
        size=2)
```

4.3 PLOTTING YOUR DATA

In Chapter 3, we saw how to use functions from the *dplyr* package to summarize our data and we produced a tibble called **RP.means** that contained the means and standard errors for *SVL.initial* for each combination of resource and predator treatments. Now, we will see how to take those summarized data and turn them into a nice looking figure. In particular, we are going to make a bar graph. Why a bar graph you ask? There are several reasons really. First, despite their ubiquity in publications, the R gurus do not like bar graphs (or barplots, as we will call them) and making one is kind of a pain in R. This is because bar graphs have an ability to hide a lot about your data (they just show the mean and whatever your error bar of choice is). But the fact that they are difficult to create makes them an excellent tool for teaching many of the ways you can, and probably should, customize your figures. That said, box plots are much more informative and are finally becoming increasingly used in published science. The second reason is that despite their downsides many people still like bar graphs and want to make them, so it is useful to know how to make one.

The most basic function to make a bar graph is **barplot()**. There are many, many arguments that can be passed to **barplot()**, which can be viewed in the help file (remember how to get to the help: *?barplot*). A slightly improved version is the function **barplot2()**, which makes plotting error bars much simpler. **barplot2()** is found in the *gplots* package. You can also make a barplot in *ggplot2*. We will go through both examples. I find it useful and instructive to demonstrate the older technique using base graphics first, as it demonstrates how you can modify every little thing in a figure in R. Much of the coding techniques are also useful in *ggplot2*, and we will use *ggplot2* for most everything in this book. If you feel confident

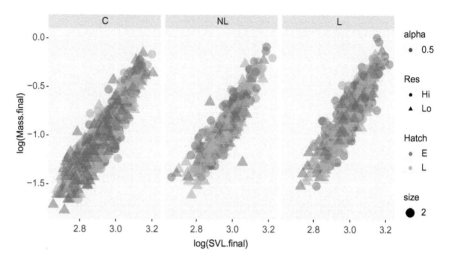

Figure 4.8 The log of the final mass of metamorphs plotted against the log of their SVLs, plotted with **qplot()**. Here, we have faceted the plot based on predator treatment, colored the points based on hatching age, and changed the shape of the points based on the resource treatment. We've also made the points 50% translucent and twice as big as normal, just for fun.

you will *never, ever, ever* use base graphics, feel free to skip ahead a few pages to the section on **ggplot2**. However, if you want to learn a little more about R and customizing figures, I encourage you to follow along with the next few pages of commands.

4.3.1 Making a barplot in base graphics

Let's start by simply plotting the bars in their most basic form. Note that I am assuming you still have the *RP.means* object we created in Chapter 3.

```
library(gplots)
barplot2(RP.means$SVL.mean)
```

We have passed to **barplot2()** the vector of mean tadpole sizes, which have been plotted in order from the first row to the last. However, this figure by itself is really awful. It has no axes titles, color, error bars, and so on (Figure 4.9).

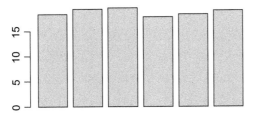

Figure 4.9 The single most boring barplot of metamorph SVL imaginable.

We can add a label to the y-axis with the argument *"ylab="* (short for "y label"). Similarly, we can use the command *"ylim="* (short for "y limits") to change the y-axis to be between 15 and 20 in order to better see the differences between our treatments. The *ylim* argument takes a vector of two numbers, the minimum and maximum of the axis you are defining. Recall from Chapter 1 that you create a vector of numbers with the concatenate function, **c()**.

The bars will be plotted left-to-right in the same order that our data frame is top-to-bottom. Thus, the first three bars are the High Resource tadpoles, and the second three bars are the Low Resource tadpoles. Those sets of bars should probably be grouped together. We can do this by changing the spacing between them, with the *"space="* argument. By default, each bar is 1 unit wide with 0.2 units between them. You can look at the bars and see that there is even spacing before the first bar and between each bar. The spacing starts at the y-axis, so when we specify the spacing, we need to think about the space before the first bar. What you pass to the *"space="* argument is a vector of numbers that defines how much space goes in between each bar. Since there are 6 bars, there are 5 gaps between them, plus the gap before the first bar, so we need a vector of 6 values.

```
barplot2(RP.means$SVL.mean,
        ylim=c(15,20), #set limits
        ylab="Mean SVL of metamorphs (mm)",  #label the y-axis
        space=c(0,0,0,0.5,0,0)) #define spaces between the bars
```

Uh oh! We changed the y-axis, but R is still trying to plot the bars all the way to 0 (Figure 4.10). We can fix that by add the argument *"xpd=F,"* which will make the bars stop at the defined y-axis minimum. Also, let's add some color. The two sets of 3 bars are the Control, Non-lethal and Lethal Predator treatments, from left to right. We can add color using the "*col=*" argument. Colors can be defined in many ways in R (see the earlier section about defining colors in R).

For this exercise, we will use prenamed colors built into R. Since these treatments represent varying types of predator treatments, we can make them increasing versions of the same basic color. Why not choose green? Note that since the progression of colors is the same in the first 3 bars as it is in the second 3 bars, we only have to define our set of colors once and they will recycle for the second set of 3 bars.

Note also that I've added one more argument in the following section to change the orientation of the y-axis scale, which makes it easier to read (the "*las=1*" part, Figure 4.11).

```
barplot2(RP.means$SVL.mean, ylim=c(15,20),
         ylab="Mean SVL of metamorphs (mm)",
         space=c(0,0,0,0.5,0,0),
         col=c("light green",
               "forest green",
               "dark green"), #set the colors
         xpd=F, #trim the plot to the visible area
         las=1) #turn the y-axis numbers 90 degrees
```

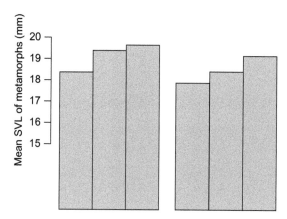

Figure 4.10 A barplot of metamorph SVL, with bars grouped by Resource treatment.

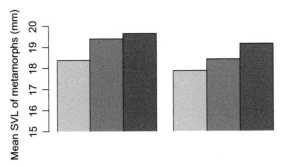

Figure 4.11 A barplot of metamorph sizes, with bars grouped by resource treatment and colors indicating the Predator treatment.

Now we should think about adding error bars. Error bars are a pain no matter what. In base R they can be added with the **arrows()** function, which is pretty stupid. In **barplot2()** we can add them inside the plot command by saying "*plot.ci=T*," as in "Plot confidence intervals? True," and then defining the location of the upper and lower limits of each error bar (simply the mean plus SE and the mean minus SE). Just like how our six bars are plotted in order from the top to the bottom of our data frame, the vector of values for the confidence intervals will follow the same rules.

```
barplot2(RP.means$SVL.mean, ylim=c(15,20),
        ylab="Mean SVL of metamorphs (mm)",
        space=c(0,0,0,0.5,0,0),
        col=c("light green",
              "forest green",
              "dark green"),
        xpd=F,
        las=1,
        plot.ci=T, #tell barplot2 to plot error bars
        ci.u=RP.means$SVL.mean+RP.means$SVL.se, #the upper limit
        ci.l=RP.means$SVL.mean-RP.means$SVL.se) #the lower limit
```

This is looking pretty good (Figure 4.12). Now, let's go ahead and add some labels to the x-axis using a function called **axis()**. This is a separate function from **barplot2()** with many different arguments that can be passed to it. Within **axis()** we will pass arguments to define: 1) which side we want the axis on (side 1 is the bottom of the plot), 2) where on the x-axis we want

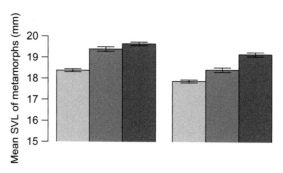

Figure 4.12 A barplot of metamorph SVL with standard error bars. Bars are grouped by resource treatment and colors indicate the predator treatment.

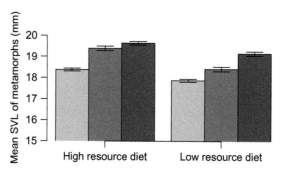

Figure 4.13 Hey look, now we have added some x-axis labels. Cool!

the labels to appear (at 1.5, and 5), and 3) what we want the labels to say (see Figure 4.13).

```
axis(side=1, #set the side
    at=c(1.5, 5), # set the location for x-axis labels
    labels=c("High resource diet",
            "Low resource diet")) #set labels
```

Lastly, we can add a legend to specify what the different colored bars indicate. **legend()** (Figure 4.14).

```
legend(x=3.5, y=20, #provide x and y coordinates for legend
    legend=c("Control",
            "Non-lethal predator",
            "Lethal predator"), #legend text
    col=c("light green",
```

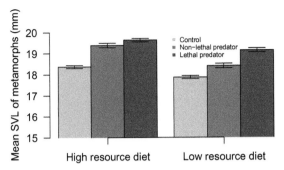

Figure 4.14 A very complete barplot of mean metamorph SVL with standard error bars. Bars are grouped by resource treatment and colors indicate the predator treatment.

```
              "forest green",
              "dark green"), #legend colors
       bty="n", #Do you want a box around the legend? Nope.
       pch=15, #the plot character to put next to the text
       cex=0.6) #the scaling of the legend, relative to 1
```

The previous **legend()** command contains arguments to define: 1) where we want the legend in x and y coordinates, 2) what we want the legend to say, 3) that we want to put squares next to the text (the "*pch=15*" part), 4) the colors of the squares (the same as our bars), 5) that we do not want to put a box around the legend (the "*bty='n'*" part), and 6) that we want the legend to be 60% of the default size (the "*cex=0.6*" part) (see Figure 4.14). It often takes a bit of trial and error to find the perfect spot for the legend. For example, if your plotting window is a different size than mine, you will likely need to move the legend a little, or you may not need to shrink the overall scaling of it. Most people get frustrated the first time they make a plot, and the legend appears on top of their bar. "How do you remove it?" they ask. You can't, but since you have presumably been working in an R script, it is really easy to just highlight all your code and rerun it to make the figure again.

Thus, in the end, we need three functions to make this figure: **barplot2()**, **axis()**, and **legend()**. That may seem cumbersome, but once you have made one, it is very easy to modify the code to make others. And you can easily change the data (the *SVL.mean* and *SVL.SE* objects) and remake the figure instantly.

Box 4.2 - Other useful tips when using base graphics for plotting

- Typing the function **box()**' at the command prompt will draw a box around your plot. Make sure to leave the parenthesis empty.
- You can add hatching to your bars with the *"density="* and *"angle="* arguments. *"density"* tells R how close together to make the lines and *angle* determines the angle to plot the lines at.
- You can shortcut the placement of the legend by just assigning the x-coordinate to be (for example) *"topright"* or *"bottomleft,"* and leaving the y-coordinate blank. Choose the basic location based on the part of your plot with the most whitespace. Look at the help function for **legend()** for all the built-in positions.

4.3.2 Making a barplot in ggplot2

Now that you've worked through all of that, let's see how you might make the same basic figure using ***ggplot2***. Whereas we used **qplot()** before, now we will use **ggplot()**. Notice that the function name does not contain the "2" on the end which the package ***ggplot2*** has. There are many aspects that are similar to building the figure in base graphics, but several that are not. I will only go over the basics of figure making with ***ggplot2*** here, but a deeper dive into the package can be found in Chapter 9.

The initial setup to make a barplot is a little more complicated in **ggplot()** than in **barplot2()** but the result is more visually appealing. Moreover, plotting with ***ggplot2*** requires a little more code than base graphics. First, we have to define where our data are found (*"data=RP.means"*). Next, we define the what are called the *aesthetics*. This is where we state which data are going on the x- and y-axes as well as which variable we want to use to fill in our bars with color. The aesthetics all go within a function called **aes()**. One of the things that can be difficult to wrap your head around when plotting with ***ggplot2*** is that the line between functions and arguments gets rather blurred. **aes()** is technically a function, but it goes within the function **ggplot()** like an argument. You can also add functions to other functions. In the following code, we will *add* a geom function to our initial plotting function. Like I said, it is a little strange at first, but it all works really well together once you get used to it.

The following first line of code just defines which variables are going to be used, but they haven't defined the *geom* to use. The next line of code, which is "added" to the initial line is where we define the geometric interpretation of the data. In **ggplot2** lingo, a vertical bar graph is actually a column graph, so you use **geom_col()**. You have to specify the *position* of the bars as *dodge*, since the default is to stack the bars upon each other. Lastly, I've added a color argument to **geom_col()** to color the borders of the bars black (Figure 4.15).

```
ggplot(data=RP.means, aes(x=Res, y=SVL.mean, fill=Pred))+
    geom_col(position="dodge", col="black") #Specify the geom
```

Hey, that was pretty cool. Our figure is actually quite complete and nice looking even in its most basic form. But let's see how you dress it up and customize it, the way we did with base graphics. The first two things we can do are to customize the color scheme and change the y-axis limits. To change the y-axis limits, we still use a "*ylim=*" argument just like before, but it is within a function called **coord_cartesian()**, which is how you change continuous axes in **ggplot2**. To change the colors, we have to adjust what is

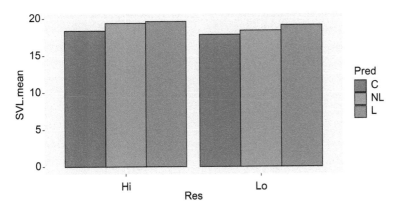

Figure 4.15 A basic barplot made with **ggplot()**. For just two lines of code, this is a nice looking figure and it even comes with a legend!

called a *scale* in **ggplot2**. Since our bars have been *filled*, we use the function **scale_fill_manual()**. If we were making a figure where we had *colored* our data, we would use **scale_color_manual()**. We assign our colors just as before, but with the "*values=*" argument (see Figure 4.16).

```
ggplot(data=RP.means, aes(x=Res, y=SVL.mean, fill=Pred))+
    geom_col(position="dodge", col="black")+ #Specify the geom
  coord_cartesian(ylim=c(15,20))+ #Adjust the y-axis limits
  scale_fill_manual(values = c("light green",
                              "forest green",
                              "dark green")) #Add color!
```

Now, we can think about adding errorbars. This is done by adding another *geom* to our figure. Yes, you can have multiple geoms in a single plot! Here, we have to define a new set of aesthetics since our error bars have plotting attributes that were not contained in our initial **aes()** function. Just like before, we have to define the minimum and maximum of the error bars, which is just the mean - standard error and mean + standard error. We have to specify that the errorbars should be "*dodged*" just like our bars, and in order to make them line up correctly you have to specify that their position is 0.9 of the full width of a bar. The reason for this is that the default width of the bars is 0.9 but the error bars are looking for bars of width 1 by default. If

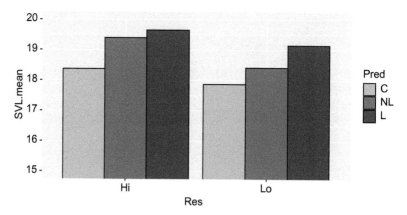

Figure 4.16 Now our figure is starting to look really good. It has custom colors and a more meaningful range of values shown on the y-axis. Notice that when we change the color scheme the legend automatically changes as well.

you want wider or skinnier bars you could add an argument to **geom_col()** to specify the width and then modify the position of the error bars to match it. We can also specify the width of the flat part of the errorbars if we want (see Figure 4.17).

```
ggplot(data=RP.means, aes(x=Res, y=SVL.mean, fill=Pred))+
    geom_col(position="dodge", col="black")+
  coord_cartesian(ylim=c(15,20))+
  scale_fill_manual(values = c("light green",
                              "forest green",
                              "dark green"))+
  geom_errorbar(aes(ymin=SVL.mean-SVL.se,
                    ymax=SVL.mean+SVL.se),
                  position=position_dodge(0.9),
                  width=0.4)#Add error bars!
```

Lastly, let's fix the legend title and the y-axis and x-axis labels. We actually customize the legend in one of the functions we've already used, **scale_fill_manual()**. The name of the legend is specified with the *"name="* argument whereas the actual factor levels are customized with the *"labels="* argument. To fix the y-axis label, we can use a function called **labs()**, wherein you specify which label you want to modify and provide the text you want for the label. We could also use this function to specify the main title for the x-axis, but it wouldn't allow us to customize the axis labels for

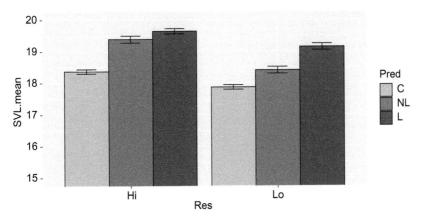

Figure 4.17 Our figure now has error bars, which makes it look very professional indeed.

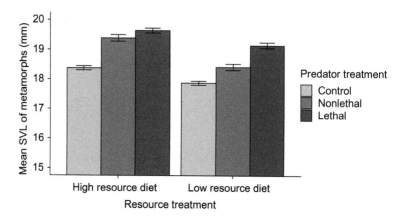

Figure 4.18 What a lovely figure! We've added customized labels all around and a nice color scheme. We've got error bars too. This is ready to send out for publication!

each group (high vs low resources). To customize those, we can use the function **scale_x_discrete()** which allows us to specify a name for the axis and the labels for each group of bars. Lastly, I've added something called a *theme* which is just a pre-defined set of graphing parameters. I personally really like **theme_cowplot()** from the, you guessed it, *cowplot* package. This theme just makes a very nice, clean figure with larger than default fonts for the axis labels (Figure 4.18).

```
ggplot(data=RP.means, aes(x=Res, y=SVL.mean, fill=Pred))+
    geom_col(position="dodge", col="black")+
  coord_cartesian(ylim=c(15,20))+
  scale_fill_manual(values = c("light green",
                               "forest green",
                               "dark green"),
                name="Predator treatment",
                labels=c("Control",
                         "Nonlethal",
                         "Lethal"))+ #Custom legend!
  geom_errorbar(aes(ymin=SVL.mean-SVL.se, ymax=SVL.mean+SVL.se),
              position=position_dodge(0.9), width=0.4)+
  labs(y="Mean SVL of metamorphs (mm)")+#Custom y-axis label
  scale_x_discrete(name="Resource treatment",
                labels=c("High resource diet",
                         "Low resource diet"))+ #Custom x-axis
  theme_cowplot() #Make it look great with cowplot!
```

⊣ Box 4.3 - Take-home message ⊢

- Whether you use base graphics or ggplot2, it is possible to make amazing, professional figures directly in R.
- Both plotting methods give you tremendous control over every little aspect of your figure.
- **ggplot2** makes it extremely easy to explore your data and really is better in just about every way imaginable compared to base graphics.

4.4 ASSIGNMENT!

Here are some things to do on your own. You can find various sample answers and code at the Github page for the book: https://github.com/jtouchon/Applied-Statistics-with-R.

1. Use **qplot()** to make a 2x3 faceted plot of age at metamorphosis against final metamorph SVL, adding informative axes labels (hint: the "*ylab*" argument also works in **qplot()**).

2. Make a barplot like the previous one, but instead of using the initial SVL data, make it from a) the final metamorph mass data and b) have it grouped by predator treatment and hatching age. Make sure it has correct error bars and give it some creative colors! Before you start coding away, think about what existing code you have in your script file that you can modify to make this figure.

Basic Statistical Analyses using R

5 Basic Statistical Analyses

The purpose of this chapter is to introduce you to some basic statistical analyses using R. This chapter assumes you are using the **RxP.clean** dataset that was created in the Chapter 3.

We will cover the following topics:

1. Assessing data normality
2. Some basic non-parametric statistics
3. Student's t-test
4. One-way Analysis of Variance (ANOVA)

5.1 DETERMINING WHAT TYPE OF ANALYSIS TO DO

We have now seen how to use a function like **qplot()** to look at your data in various ways. For example, you can plot a histogram of your response variable and see how it is distributed. However, as you move beyond the initial steps of data exploration and start to think about data analysis, there are several questions you should ask yourself. The most important of which is just *what kind of data do you have?*

Applied Statistics with R: A Practical Guide for the Life Sciences. Justin C. Touchon, Oxford University Press (2021). © Justin C. Touchon. DOI: 10.1093/oso/9780198869979.003.0005

This might seem like a simple question at first, but it is paramount for determining the analyses you will conduct. When we talk about data, are we talking about your response or your predictor variables? The answer is both of course. Knowing *the shape* of your response and predictor variables will determine what sort of analysis you do. In addition to simply knowing if the data are *normal* or not, you should be mindful of if you have one predictor or multiple predictors, and if your predictors are continuous (a bunch of numbers) or discrete (different categories). For this chapter and the next, we will just concern ourselves with analyzing data where the response variable is normally distributed.

Box 5.1 - What are "normal" data?

A great place to start of course is by considering what do we even mean with the term "normal." Many standard statistics rely on data being approximately *normally distributed*. These statistical models are also called *parametric statistics*. But, what does it mean to say that data are "normal?" Normal data are those that have an even spread of values above and below the mean value and those data are clustered such that the majority of values fall close to the mean. Normal data can be any real value, positive or negative. If you look at Figure 5.1, you can see a visual depiction of a normal distribution. The vertical lines represent *standard deviations* above and below the mean and the percentages represent the percentage of values that are expected to fall within each standard deviation. "Normal" is an error distribution, also knows as a Gaussian distribution (more on error distributions in a bit). Here is some useful terminology if this is sounding totally foreign to you.

- Data is plural, it usually refers a bunch of numbers. The singular term for a single number would be datum.

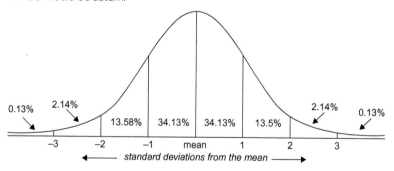

Figure 5.1 A Gaussian, or normal, distribution showing the even spread of values around the mean.

- Mean is a singular term. It refers to a single value which describes something about a bunch of numbers. It is the sum of all the numbers divided by how many of them you have.
- The variance is the average of how different each value in the dataset is from the mean of the dataset, squared.
- The standard deviation is calculated as the square root of the variance of a set of data.

If your data are normally distributed, you would expect 61.26% of all the values to fall within one standard deviation of the mean. Similarly, you would expect 95.42% to fall within two standard deviations. The total of the area under the curve is of course 100%.

In R, nearly all basic models that assume normality of your data are coded with the same basic design.

```
lm(Response ~ Predictor1 + Predictor2 +..., data)
```

The previous code indicates 1) we have a function called **lm()** (short for "linear model") that specifies what type of analysis we are going to do, 2) we use a "~" to separate the response (aka dependent variable) written on the left side of the equation from one or multiple predictors (aka independent variables) written on the right side of the equation, and 3) we specify where R will find the data. This syntax might seem familiar, as we used it both to create basic plots of data and when using the **aggregate()** function to summarize our data in Chapter 3. The following table shows how different combinations of continuous and discrete predictor variables determine which type of model you might do.

Table 5.1 All the different flavors of linear model you can do

Data Distribution	Response Variable	Predictor Variable	# Predictors	Test
Normal	continuous	discrete	one	One-way ANOVA
Normal	continuous	discrete	multiple	Multi-way ANOVA
Normal	continuous	continuous	one	Linear regression
Normal	continuous	continuous	multiple	Multiple regression
Normal	continuous	discrete & continuous	multiple	ANCOVA

Analysis of Variance (ANOVA), Analysis of Covariance (ANCOVA)

The great thing about R is that as long as your response variable is normally distributed, you can code any model using the same syntax! The only important thing is what your predictor variable(s) look like.

Box 5.2 - Reminder: The Data

This section serves as a reminder of the RxP experiment data, which come from the article: Touchon, J.C., McCoy, M.W., Vonesh, J.R., and Warkentin, K.W. 2013. "Effects of hatching plasticity carry over through metamorphosis in red-eyed treefrogs." *Ecology.* 94(4): 850–60.

The experiment consisted of 96 400 L mesocosm tanks arrayed in an open field in the northwest corner of Gamboa, Panama. The mesocosms were spatially arranged in 8 blocks of 12 tanks each. Each block consisted of 1 tank from each of 12 unique treatment combinations (hatching age, predator treatment, and resource level). Each tank began with 50 tadpoles and the experiment ended when all tadpoles reached metamorphosis or had died. Recall that we wanted to know how the hatching age of a frog embryo might affect its development to metamorphosis under various combinations of predators and resources.

Predictor variables

- Hatching age: Early (4 days post-oviposition) or Late (6 days post-oviposition)
- Predators: Control, Nonlethal (caged) dragonfly larvae, or Lethal (free-swimming) dragonfly larvae
- Resources: Low (0.75 g) or High (1.5 g) food level, added every 5 days
- Block where the metamorph was reared
- Tank where the metamorph was reared

Response variables

- Age at hatching, both in terms of time since eggs were oviposited and time since emergence began (defined as Day 1)
- Snout-vent (SVL) length at emergence
- Tail length at emergence
- SVL at completion of tail resorption
- Mass at completion of tail resorption
- Number of days needed for each metamorph to fully resorb the tail

During the course of the experiment disease broke out in 18 of the mesocosms containing Nonlethal predators and thus those tanks have been excluded from the dataset.

In addition, several outliers were removed in Chapter 3. The analyses conducted from here until the end of the book assume that you have already been through that chapter and have removed the outliers and are using the RxP.clean dataset.

5.2 AVOIDING PSEUDOREPLICATION

Although we have data on nearly 2500 individual metamorphs, those data are not all independent from one another. This is because groups of tadpoles were raised in common environments (the mesocosms, aka tanks) and variation between tanks may make certain individuals more similar than others. If we treat all individuals as independent, we are committing pseudoreplication, which is when you artificially inflate your sample size of independent observations. This was first mentioned in Chapter 2, but for more details, see the classic article Hurlbert, S.H. 1984. "Pseudoreplication and the design of ecological field experiments." *Ecological Monographs* 54(2): 187–211. It's a little long, but it's a great read!

One easy way to avoid pseudoreplication is to utilize the mean value for each tank instead of that of individuals. We can summarize our entire dataset with relative ease using the **summarize()** function that was introduced in Chapter 3. We will have to get rid of column 1 on Individual ID and column 3 which shows the tank within each block because those would not be meaningful to average at the tank level. Recall that in order to use **summarize()** we first: 1) define our dataset, then 2) define the different variables we want to group our data by with the **group_by()** function, and lastly 3) use **summarize()** to create the variables that are the mean values from the raw dataset. At each of the three steps you pipe your data from one line to the next. You could also use the **aggregate()** function by defining the 7 columns to summarize (our response variables) and binding them together in a single object using the function **cbind()**, which *binds* two or more columns together, then give it the various predictor variables we want to use to summarize the data. Here, we will use **summarize()**, which just makes more sense. Recall that these functions are found in the *dplyr* package.

One other thing to notice in the following code is that there are more variables than necessary in the **group_by()** function. All we need to uniquely identify each tank is the variable *"Tank.Unique"* or the three treatments in combination. Including all of them, as well as the

"*Block*" variable, we will merely carry more columns through to our new summarized dataset. Lastly, notice that we are storing the output of this set of code as a new object called "*RxP.byTank*."

```
library(dplyr)
RxP.byTank<-RxP.clean %>%
            group_by(Block,Tank.Unique,Pred,Hatch,Res) %>%
            summarize(Age.DPO = mean(Age.DPO),
                      Age.FromEmergence = mean(Age.FromEmergence),
                      SVL.initial = mean(SVL.initial),
                      Tail.initial = mean(Tail.initial),
                      SVL.final = mean(SVL.final),
                      Mass.final = mean(Mass.final),
                      Resorb.days = mean(Resorb.days))
RxP.byTank<-as.data.frame(RxP.byTank)
str(RxP.byTank)
```

```
## 'data.frame':    78 obs. of  12 variables:
##  $ Block            : int  1 1 1 1 1 1 1 1 1 1 ...
##  $ Tank.Unique      : int  1 2 3 4 5 6 7 8 9 10 ...
##  $ Pred             : Factor w/ 3 levels "C","NL","L": 2 1 1 .
##  $ Hatch            : Factor w/ 2 levels "E","L": 2 1 2 2 1 ...
##  $ Res              : Factor w/ 2 levels "Hi","Lo": 1 1 1 2 ...
##  $ Age.DPO          : num  47.2 45.4 53.8 56.9 64.8 ...
##  $ Age.FromEmergence: num  13.2 11.4 19.8 22.9 30.8 ...
##  $ SVL.initial      : num  19.4 18.4 18.9 18.8 19.7 ...
##  $ Tail.initial     : num  4.83 5.37 4.8 4.63 5.43 ...
##  $ SVL.final        : num  19.7 19 19.1 19.1 20.1 ...
##  $ Mass.final       : num  0.418 0.382 0.412 0.382 0.486 ...
##  $ Resorb.days      : num  3.49 3.79 3.51 3.65 4.22 ...
```

Summarizing our data in this way is nice because now our observations are independent (although we might want to think about the fact that tanks come from within blocks). However, we have reduced our dataset of nearly 2500 individuals to just 78 observations! That is throwing out or ignoring a tremendous amount of data that we worked very hard to obtain. In Chapter 8 on Mixed Effects Models, we will discuss how you can utilize all of your data while controlling for the non-independence of certain observations.

Box 5.3 - Saving your hard work out of R

Now that you've put in the effort to summarize your dataset (which you will use for the next few chapters!) you might think it is a good idea to save it back out of R to your hard drive or the cloud, thereby allowing you to view it in another program, email it to a collaborator, or import it at a future point in time without re-running all your code again. Just like when you used the function **read.csv()** to read in your data, you can use its parallel function **write.csv()** to get your data back out of R and saved as a file. All you have to provide is one argument: the R object you want to save. You can (and probably should) also specify a path (in quotes) for where you want the file to be saved, but if you don't R will save it to your working directory. I would also recommend a third argument, which specifies that you don't want R to create a new column that lists the row numbers of your data frame (indicated with *"row.names=FALSE"*). For example, if you are working on a Mac and want to save RxP.byTank as a file on your desktop, you would type the following:

```
#This will create a .csv file on my desktop called RxP_
   byTank.csv!
write.csv(RxP.byTank, "~/Desktop/RxP_byTank.csv",
   row.names=FALSE)
```

Recall from Chapter 1 that it is a good idea to have a general naming convention for your variables, data objects, and so on. My personal preference is to use the underscore (_) to create spaces in actual file names on my computer and to generally use a period (.) to create spaces within R. You might have a different preference, which is perfectly fine. But, in the previous code, I've saved the data frame *"RxP.clean"* as a file called *"RxP_clean.csv."*

Note that when you read the file back in, any factors will be back in the default alphabetical order (i.e., Pred will be ordered C, L, NL instead of C, NL, L as we changed it to in Chapter 3).

5.3 TESTING FOR NORMALITY IN YOUR DATA

How do you know if your data fit a normal distribution? That is an essential but somewhat difficult question to answer. There are multiple ways to try and address that question. The first and most important thing to do is simply to plot a histogram of your data—to look at the spread of values. In addition, we will discuss two different approaches: 1) normality tests, and 2) using maximum likelihood to estimate the most appropriate error distribution.

5.3.1 Looking at the data

The first and easiest thing to do if you want to see if your data are normal or not is to plot the histogram of values. You could use either the **hist()** function or the **qplot()** function found in the ***ggplot2*** package. Either way, a histogram should give you a general sense of what your data look like. Remember that data which are normally distributed will have a relatively even spread of values above and below the mean, like that shown in Figure 5.1. Let's look at the variables "*Age.FromEmergence*" and "*SVL.final*" and plot them using **qplot()** from the ***ggplot2*** package. Remember, you have to load the package first using the **library()** function. Notice in the following code that we can use the "*bins=*" argument to specify how fine we want the histogram to break up the data. Also notice I've loaded the package ***cowplot*** which contains the function **plot_grid()** which allows us to plot multiple different figures together in a single window.

```
library(ggplot2)
library(cowplot)
a<-qplot(data=RxP.byTank,
         x=SVL.final,
         geom="histogram",
         bins=8)
b<-qplot(data=RxP.byTank,
         x=Age.FromEmergence,
         geom="histogram",
         bins=10)
plot_grid(a,b,ncol=2)
```

Looking at these two histograms (Figure 5.2) is quite informative. The SVL data appear almost normal, although the data have a slight tail to the right. The Age data are, however, very skewed; most individuals emerged very early in the period of metamorphosis, but the tail is very long with individuals still emerging from tanks more than 100 days after the first metamorphs. Many biological patterns can be made normal with log-transformation, so a good first step is to see what effect that has on our data. Data that can be made normal upon log-transformation are referred to as "lognormal." As we explored in Chapter 3, plotting data as a density plot can also be useful, and so both will now be shown in Figure 5.3.

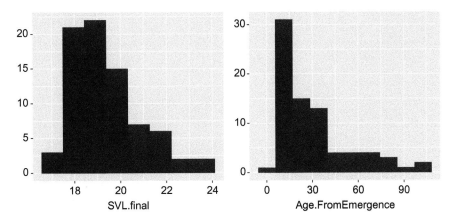

Figure 5.2 Histograms of SVL.final and Age.FromEmergence for *A. callidryas* metamorphs.

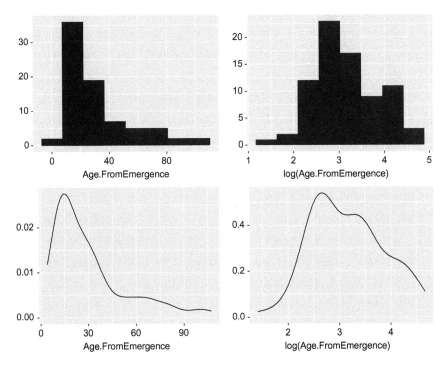

Figure 5.3 Histograms and density plots of Age.FromEmergence and log-transformed Age.FromEmergence for *A. callidryas* metamorphs.

```
#Make histograms of the original and log-transformed data
a<-qplot(data=RxP.byTank,
         x=Age.FromEmergence,
         geom="histogram",
         bins=8)
b<-qplot(data=RxP.byTank,
         x=log(Age.FromEmergence),
         geom="histogram",
         bins=8)
#Make density plots of the original and log-transformed data
c<-qplot(data=RxP.byTank,
         x=Age.FromEmergence,
         geom="density")
d<-qplot(data=RxP.byTank,
         x=log(Age.FromEmergence),
         geom="density")
#Plot them all in a 2x2 grid
plot_grid(a,b,c,d, ncol=2, nrow=2)
```

Log-transformation definitely appears to have helped normalize the data, which can be seen in both the histograms and the density plots (Figure 5.3). Where the raw data have a peak around 20 days and a very long tail out to the right, the log-transformed data have a relatively even spread of values around the peak, which is a little below 3 (on a log-scale). These figures imply that maybe our data are indeed lognormally distributed. We can use a normality test to measure this effect statistically.

5.3.2 Normality tests

A classic way to identify if your data are suitable for parametric statistics is through the use of normality tests. These tests compare your dataset to a normal distribution with a similar mean and variance and assign a p-value of the probability that your data are normal. Smaller p-values (e.g., $p < 0.05$) indicate that the data are less likely to be normal. This is an important distinction: a small p-value does not mean your data "are not normal," it means that it is unlikely they come from a larger pool of data that are normally distributed. The most widely used normality test is the Shapiro-Wilk test, executed in R using the function **shapiro.test()**. This is an old function in R and does not accept the "*data=*" argument, so you have to just specify the vector of data you are interested in. Recall that you specify a particular column in a data frame with the "$" operator (i.e., "df$column").

```
shapiro.test(RxP.byTank$Age.FromEmergence)
```

```
##
##   Shapiro-Wilk normality test
##
## data:   RxP.byTank$Age.FromEmergence
## W = 0.82366, p-value = 3.143e-08
```

```
shapiro.test(RxP.byTank$SVL.final)
```

```
##
##   Shapiro-Wilk normality test
##
## data:   RxP.byTank$SVL.final
## W = 0.93731, p-value = 0.0008234
```

The Shapiro-Wilk test indicates that neither of these variables are normally distributed. However, what about when we log-transform them, as we did above?

```
shapiro.test(log(RxP.byTank$Age.FromEmergence))
```

```
##
##   Shapiro-Wilk normality test
##
## data:   log(RxP.byTank$Age.FromEmergence)
## W = 0.97304, p-value = 0.09666
```

```
shapiro.test(log(RxP.byTank$SVL.final))
```

```
##
##   Shapiro-Wilk normality test
```

```
##
## data:   log(RxP.byTank$SVL.final)
## W = 0.95421, p-value = 0.006943
```

Log-transformation of the tank-summarized data normalizes Age.From Emergence and greatly improves the normality of SVL.final. Although the p-value is still substantially smaller than 0.05, clearly log-transformation has helped to normalize the SVL.final data. There are several other normality tests found within the ***nortest*** package if you wish to explore them.

Box 5.4 - How to estimate the most appropriate error distribution using maximum likelihood

Data in the real world are often messy, that is a fact of life for most scientists. This means that data often do not fall exactly within the confines of certain data classifications, such as "normal" vs "non-normal." An alternative to normality tests is to simply estimate which established error distribution most accurately reflects your data. Thus, instead of trying to define what kind of data you have, you are instead trying to understand what your data are closest to. This can be done using the **fitdistr()** function (short for "fit distribution") found within the ***MASS*** package.

5.3.3 Understanding error distributions

I've thrown around the term "error distribution" a lot already, but what is an error distribution you ask? In essence, the "error" is a way of describing the shape of the data surrounding a mean. The "error" does not mean mistakes, but rather it refers the difference between your collected data and some, probably unknown or unknowable, true population mean and the spread of data around that mean. As we saw earlier, "normal" is a type of error distribution. It describes a set of numbers where the spread of values is relatively even above and below the mean of the values. There are many different error distributions and we will explore them more in Chapter 7. If your data are not normally distributed, do not fret! There is a whole universe of models that are designed to handle non-normal data. I will also show you some old-school ways to analyze non-normal data in just a few pages.

The **fitdistr()** function uses maximum likelihood to assess the fit of your data against a predefined error distribution and assigns an Akaike information criterion (AIC) score for the fit. This allows you to compare the fit of different error distributions using AIC. The full list of possible error distributions can be found in the help file for **fitdistr()**. What is AIC you ask? AIC stands for Akaike's Information Criterion and, in a nutshell, it is a measure of how well a model fits, i.e., how well a model explains the variance in the data. Here are three important things to know about AIC scores. 1) AIC scores are only comparable on the *exact* same response variable. 2) Although AIC scores are unitless, smaller AIC scores are better. 3) AIC values within 2 of one another are generally considered to be pretty equivalent. This is a pretty simplistic take on a complex topic, but for most purposes it works.

Let's continue to explore the SVL.final and Age.FromEmergence variables. We will create a new object for each error distribution using **fitdistr()** and compare those objects with the **AIC()** function. You will need to load the **MASS** package into the active memory using the **library()** command. **fitdistr()** takes just two arguments: the vector of data you are examining and the distribution you want to compare it to. In the example below, I've made two objects called simply "*fit1*" and "*fit2*" and each utilize the same data, the SVL.final column from the RxP.byTank dataset. In "*fit1*," I've specified to compare the data to a normal distribution and in "*fit2*" to a lognormal distribution.

```
library(MASS)
fit1<-fitdistr(RxP.byTank$SVL.final, "normal")
fit2<-fitdistr(RxP.byTank$SVL.final, "lognormal")
AIC(fit1,fit2)
```

```
##        df      AIC
## fit1    2  280.7438
## fit2    2  276.4907
```

The AIC scores indicate that as suspected, the lognormal distribution is a better fit to the SVL.final data than a normal error distribution. Similar to what we saw with the histograms, the effect is not huge, but lognormal is slightly better. The original data on Age.FromEmergence are integer values and the long tail indicates that the data might be Poisson or even negative binomially distributed (if these terms are like gibberish to you, just wait until Chapter 7). Let's also see if a lognormal distribution fits the data better than a normal distribution, and for now, let's just worry about those two distributions.

```
fit1<-fitdistr(round(RxP.byTank$Age.FromEmergence), "normal")
fit2<-fitdistr(round(RxP.byTank$Age.FromEmergence), "lognormal")
AIC(fit1,fit2)
```

```
##        df       AIC
## fit1    2  718.7014
## fit2    2  665.9925
```

For Age.FromEmergence, the AIC scores once again indicate that the lognormal distribution is the best fitting error distribution. The effect appears to be larger than for SVL.final too, which is similar to what we saw with the Shapiro-Wilk test.

To make our lives easier, let's go ahead and create new variables in the RxP.byTank dataset that are the log-transformed values of SVL.final and Age.FromEmergence. To add new variables to an existing data frame, simply create a new object that refers to the data frame and R will automatically tack it on the end. While we are at it, we might as well log-transform Age.DPO as well.

```
RxP.byTank$log.SVL.final<-log(RxP.byTank$SVL.final)
RxP.byTank$log.Age.FromEmergence<-log(RxP.byTank$Age.
    FromEmergence)
RxP.byTank$log.Age.DPO<-log(RxP.byTank$Age.DPO)
```

You can confirm that this worked with the **str()** function.

```
str(RxP.byTank)
```

```
## 'data.frame':      78 obs. of   15 variables:
##  $ Block                : int  1 1 1 1 1 1 1 1 1 ...
##  $ Tank.Unique          : int  1 2 3 4 5 6 7 8 9 10 ...
##  $ Pred                 : Factor w/ 3 levels "C","NL","L": 2 .
##  $ Hatch                : Factor w/ 2 levels "E","L": 2 1 2 ..
##  $ Res                  : Factor w/ 2 levels "Hi","Lo": 1 1 ..
##  $ Age.DPO              : num  47.2 45.4 53.8 56.9 64.8 ...
##  $ Age.FromEmergence    : num  13.2 11.4 19.8 22.9 30.8 ...
##  $ SVL.initial          : num  19.4 18.4 18.9 18.8 19.7 ...
##  $ Tail.initial         : num  4.83 5.37 4.8 4.63 5.43 ...
##  $ SVL.final            : num  19.7 19 19.1 19.1 20.1 ...
##  $ Mass.final           : num  0.418 0.382 0.412 0.382 ...
##  $ Resorb.days          : num  3.49 3.79 3.51 3.65 4.22 ...
##  $ log.SVL.final        : num  2.98 2.94 2.95 2.95 3 ...
##  $ log.Age.FromEmergence: num  2.58 2.43 2.99 3.13 3.43 ...
##  $ log.Age.DPO          : num  3.85 3.82 3.99 4.04 4.17 ...
```

5.4 NON-PARAMETRIC TESTS

Back in the day, if your data were not normally distributed you would likely have resorted to using what are called non-parametric statistics or unnecessary forms of data transformation (more about that in Chapter 7). Statistics that rely on normally distributed data are called "parametric." Most people would now agree that these sorts of techniques are generally outdated and not necessary. That said, there may be times when you find them useful and they are good to know about.

Let's start by considering why non-parametric statistics were, and might still be under some conditions, considered useful. The primary reason is that non-parametric statistics make no assumptions about your data. You can pretty much have ANY kind of data and still be able to analyze them with non-parametric stats. Crazy right? They don't need to be normal; they don't need to be anything. And therein lies their beauty. When all else fails, you can use non-parametric statistics.

Now let's discuss why non-parametric statistics are frowned up these days. For one thing, you are not actually analyzing your data any more. Seriously, you don't analyze your actual data. Rather, you are usually testing the difference in the **ranked order** of your data. What does that mean? Well, imagine you take all the data you have painstakingly collected and line the numbers up from smallest to biggest (i.e., order them), then assign them ranks (e.g., 1, 2, 3, etc.). When it comes time to run the analysis, you are no longer analyzing the numbers themselves (i.e., your data), but the ranks of them. It's like asking "does group A have more big numbers than group B?" as opposed to asking if the values in group A are actually different from those in group B. More importantly, non-parametric tests are generally considered to have **low power**, meaning that their ability to distinguish between two groups that are not **extremely** different is not very good. Presumably you want statistics that can discriminate the differences between your treatment groups, and non-parametric statistics can only really do that if the differences are huge. The good thing is that that means non-parametric statistics are very conservative and are unlikely to give you a false positive result. If you get a significant result using these methods, you can be pretty confident that your groups are indeed different. At the same time, if you get a significant difference using these techniques you probably knew there was a difference already!

5.4.1 The Mann-Whitney U test and the Kruskal-Wallis test

I will discuss two non-parametric tests. There are more of course and if you desire to look them up, you can do so on your own time. The Mann-Whitney U test is what you would you use when you want to compare two independent samples of data. It is essentially a non-parametric version of the t-test (described later in this chapter). Similarly, the Kruskal-Wallis test is used when you have more than two groups of data, so it is a non-parametric version of an ANOVA (also described later). If you care to look them up, a Wilcoxon signed-rank test is a non-parametric version of a paired t-test and a Friedman test is a non-parametric version of a repeated measures ANOVA.

We saw earlier that even after log-transformation, the data on the final SVL of metamorphs is not normally distributed. Why bother log-transforming it when we can just analyze it with non-parametric tests?

If we want to see if there is an effect of resource treatment on SVL, we could use a Mann-Whitney U test. Remember, this isn't actually asking if the means of the SVLs of metamorphs from the two resource treatments differ, it is asking if the ranks of the values differ.

Now, the most confusing thing about conducting these analyses in R is that the function to do a Mann-Whitney U test (remember, that is like an independent samples t-test) and a Wilcoxon signed-rank test (like a paired samples t-test) is that they use the same function, which is **wilcox.test()**. If you want to do an actual Wilcoxon signed-rank test, you just add an argument to indicate "*paired=T*".

```
#Remember, despite the name of the function
#this is doing a Mann-Whitney U test!!
wilcox.test(SVL.final~Res, data=RxP.byTank)
```

```
##
##   Wilcoxon rank sum exact test
##
## data:  SVL.final by Res
## W = 964, p-value = 0.04204
## alternative hypothesis: true location shift is not equal to 0
```

Similarly, if we wanted to test for differences in our predator treatment, which has three levels, we can run a Kruskal-Wallis test, which is analogous to a one-way ANOVA.

```
kruskal.test(SVL.final~Pred, data=RxP.byTank)
```

```
##
##   Kruskal-Wallis rank sum test
##
## data:  SVL.final by Pred
## Kruskal-Wallis chi-squared = 32.505, df = 2, p-value = 8.744e-08
```

Both of these non-parametric tests indicate significant differences based on our treatments, so we can probably be pretty confident that resources and predators did actually affect the SVL of *A. callidryas* froglets at metamorphosis. What about for hatching age? Go on, run the model, you have the skills. In the following chapters we will see how to analyze these data with more sophisticated statistics that properly incorporate the error distribution of the data.

Box 5.5 - Who was Student and what was his t-test?

One of the simplest *parametric* statistical analyses to conduct is Student's t-test. The test is not designed just for students, but rather the name "Student" was used as a pseudonym by William Sealy Gosset, who invented the test. The backstory of why and how Gosset created the test is very interesting and worth a read: https://en.wikipedia.org/wiki/Student%27s_t-test#History). The quick version is that Gosset worked for the Guiness brewing company in Ireland and wanted to compare the production of different fields of barley (one of the key ingredients in beer), so he designed and published a statistical test designed specifically for small sample sizes.

There are multiple uses for the t-test, and it is designed for times when you have fewer than 30 samples. With a large sample size, the t-test is equivalent to a linear model (more on that later). With a t-test, you can:

1. Compare the means of two independent groups of normally distributed data.
2. Compare the means of two groups of paired data (e.g., before and after measurements).
3. Compare the mean of a single set of data with a hypothetical mean (e.g., from a previous study).

5.4.2 Student's t-test

The function to conduct a t-test in R is simply **t.test()**. Although there are several ways to code the model, the most straightforward is just **t.test(response~predictor)**. For example, maybe we would like to know if there is a significant difference in the time to metamorphosis (Age.FromEmergence) based on the resource treatment tadpoles experienced. One might understandably expect that tadpoles fed more would grow to a larger size (duh!). In the following code, we don't call each variable directly from the data frame using the $ operator, but instead we use "*data=*" to specify where the model should look for any variables we give it, as was done above for non-parametric statistics.

Note that by default R will assume that the variances in your two groups are unequal and will adjust the degrees of freedom for your t-test accordingly. This can be confusing for some folks because it will give slightly different results than what you might calculate by hand. If you are reasonably sure the variances are similar, add the argument "*var.equal=T*," which will conduct a straightforward t-test like you were taught in school. If you suspect the variances are unequal in your two groups, simply leave it out.

```
t.test(log.Age.FromEmergence~Res, data=RxP.byTank, var.equal=T)
```

```
##
##   Two Sample t-test
##
## data:   log.Age.FromEmergence by Res
## t = -4.6246, df = 76, p-value = 1.511e-05
## alternative hypothesis: true difference in means is not
## equal to 0
## 95 percent confidence interval:
##   -0.9493877 -0.3778055
## sample estimates:
## mean in group Hi mean in group Lo
##           2.835664          3.499261
```

What does all of that output tell us? First, R tells us what we did (a two-sample t-test) and reminds us of what we are analyzing, the log.Age.FromEmergence variable by the resource treatment. Next, we get the t-statistic, the degrees of freedom, and the p-value for the test. The p-value tells us the probability of the null hypothesis, written as H_0, which states that the difference in the means between our two groups are not different (the flip side is the shown H_1, or alternative hypothesis, that the difference in the means is not 0). Since the p-value is much smaller than 0.05 (it's 0.00001511 to be exact), by convention we take this as evidence that *it is quite probable* that the difference in the means is not 0. That's a weird way to say, "they are different," since a difference of 0 is not a difference. Most people would just say "they are significantly different," which is technically incorrect, but practically correct in most cases. The 95% confidence interval is the upper and lower limits on what the model estimates as the difference in the means. You can see that it does not span 0, and thus supports our belief that there is a large difference between the two groups. If the confidence interval included 0, you wouldn't be too confident that your groups were different, would you?

Thus, we can conclude that the mean age at emergence (in terms of time since the first metamorph emerged) differed if the rearing tank received high or low resource levels. The last two lines of the output tell us the means of each group. Since we log-transformed the data, we can use the **exp()** function to exponentiate the values back into real values, which are in days.

```
exp(2.835664)#mean of the Hi treatment
```

```
## [1] 17.04171
```

```
exp(3.499261)#mean of the Lo treatment
```

```
## [1] 33.09099
```

So, the means are very different; metamorphs from the High resource tanks emerged in nearly half the time of those from the Low resource tanks. How do you think this would have been different if we had not averaged the values for each tank? We can similarly test if Age.FromEmergence differs, for example, based on Hatching treatment.

```
t.test(log.Age.FromEmergence~Hatch, data=RxP.byTank, var.equal=T)
```

```
##
##  Two Sample t-test
##
## data:  log.Age.FromEmergence by Hatch
## t = -1.0807, df = 76, p-value = 0.2832
```

```
## alternative hypothesis: true difference in means is not
## equal to 0
## 95 percent confidence interval:
##  -0.4952698  0.1468414
## sample estimates:
## mean in group E mean in group L
##          3.080356          3.254570
```

```
#Means of Early and Late hatching ages
exp(3.080356)
```

```
## [1] 21.76615
```

```
exp(3.254570)
```

```
## [1] 25.90847
```

It looks like Hatching age did not appear to have as strong effect on age at metamorphosis (just a 4-day difference), particularly in comparison to the other treatment variables we have looked at.

We can visualize these differences very easily with boxplots (Figure 5.4).

```
res.plot<-qplot(data=RxP.byTank,
                x=Res,
                y=log.Age.FromEmergence,
                geom="boxplot",
                fill=Res)
hatch.plot<-qplot(data=RxP.byTank,
                  x=Hatch,
                  y=log.Age.FromEmergence,
                  geom="boxplot",
                  fill=Hatch)
plot_grid(res.plot, hatch.plot, ncol=2)
```

One thing we should think about is if it is even appropriate to use the t-test in this scenario. As mentioned above, with a large enough sample size a t-test becomes numerically identical to a linear model. The linear model is the common framework for conducting models commonly referred to as ANOVA, ANCOVA, and linear regression. Since we have 78 tanks (i.e., 78

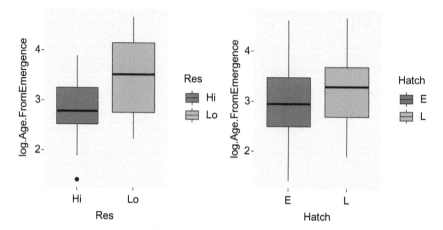

Figure 5.4 We can clearly see that the difference in resource treatments was very large, whereas the difference in hatching treatments was pretty small.

independent observations) our sample size is certainly large enough to warrant using a linear model.

5.5 INTRODUCING LINEAR MODELS

The model commonly referred to as a linear model, or **lm()**, is one of the most flexible and useful in all of statistics. This flexibility leads the model to also be known as the *general* linear model to some. This is a bit confusing with the *generalized* linear model which we will discuss in Chapter 7, and so we will just use R's preferred lingo, the linear model. The constraint on a linear model is that the response variable must be normally distributed, but the predictor variable (or variables) can be either continuous or discrete (i.e., categorical). See the table at the start of this chapter for the lay terms which are commonly applied to linear models with various combinations of predictor variables.

Box 5.6 - Useful functions for linear models

There are several functions that are very useful for any kind of linear model. These include the following:

- **summary()**—Summarizes your model and gives you important information such as adjusted R-squared and treatment means, as well as the F-statistic and p-value for the model. For linear regression and ANCOVA this also provides the slope and intercept for the line of best fit.

- **anova()**—Provides a very brief summary of the overall model but does not provide information on individual levels within factors. The statistical significance of each predictor is calculated in a stepwise manner, adding each factor one-by-one.
- **Anova()**—Provides similar summary information to the **anova()** function, but here statistical significance of each predictor is calculated assuming all other factors are in the model. This function is found within the *car* package and is completely identical to **anova()** for most models. However, I personally like **Anova()** better than **anova()** for calculating summary statistics because you know they are always correct and are more conservative.
- **plot()**—Provides diagnostic plots of the residuals of the model, which are useful for assessing model fit and balance.

5.6 ONE-WAY ANALYSIS OF VARIANCE—ANOVA

One of the most common types of statistical analyses is the Analysis of Variance, called ANOVA for short. In its simplest form, an ANOVA is a statistical test of whether or not the means of multiple (more than two) groups are equal. Thus, it is essentially the same as the t-test but for more than two groups (in fact, you can use this to do the exact same hypothesis test as the t-test). The resulting p-value and test statistic give us some idea of the probability that those group means are indeed different. For example, perhaps we want to know if log.Age.FromEmergence differs across our three predator treatments (Control, Nonlethal, and Lethal). Here, we will actually make our model into an R object and then examine it using the **summary()** function. Note that we could have done that in the t-test example, we just chose not to. Also, remember that we are using the log-transformed version of your variable so that it is normally distributed.

```
lm1<-lm(log.Age.FromEmergence~Pred, data=RxP.byTank)
summary(lm1)
```

```
##
## Call:
## lm(formula = log.Age.FromEmergence ~ Pred, data = RxP.byTank)
##
## Residuals:
##       Min       1Q    Median       3Q       Max
```

```
## -1.85101 -0.34088 -0.07498   0.37065   1.40721
##
## Coefficients:
##              Estimate Std. Error t value Pr(>|t|)
## (Intercept)    3.5325     0.1107  31.920  < 2e-16 ***
## PredNL        -0.2677     0.2006  -1.335    0.186
## PredL         -0.7725     0.1565  -4.936 4.68e-06 ***
## ---
## Signif. codes:  0 '***' 0.001 '**' 0.01 '*' 0.05 '.'
      0.1 ' ' 1
##
## Residual standard error: 0.626 on 75 degrees of freedom
## Multiple R-squared:  0.2483, Adjusted R-squared:  0.2283
## F-statistic: 12.39 on 2 and 75 DF,  p-value: 2.244e-05
```

So, what all does that mean? We created an object called "*lm1*" and then used the **summary()** function to look at it. The top of the summary output tells us the model we made and gives us information about the distribution of the residuals around the mean. Residuals are how different each of the actual data points are from the predicted fit from the model. Thus, if our data fit the model well, we would expect roughly an even spread of residuals above and below the median, which should be close to zero. The section entitled "*Coefficients*" provides information about the estimated mean for each level (C, NL, and L) within the factor (Pred). The way R reports estimates of means is that a baseline is established alphabetically, and everything else is in relation to that baseline. Thus, in the output for **lm1**, what is labeled as "*(Intercept)*" is the mean for the Control treatment (because C comes before NL or L). Note that coefficients will be reported in alphabetical order by default but if you have changed the order of your factor levels, as we've done here, that is how they will get reported. All of the values reported are *on the scale of your response variable*. In this case, that means the numbers are log-transformed and to make sense of them we have to exponentiate them. Thus, the age of metamorphs from the Control tanks is:

```
exp(3.5325)
```

```
## [1] 34.20938
```

or 34.21 days. The subsequent "*Estimates*" are how much those treatments differ from the baseline. Thus, the "*PredNL*" level is 0.2677 fewer days than the Control treatment on the log scale, or

```
exp(3.5325 - 0.2677)
```

```
## [1] 26.17488
```

26.17 days. Lastly, the logged age at metamorphosis for the L treatment is the last line, and is

```
exp(3.5325 - 0.7725)
```

```
## [1] 15.79984
```

or 15.80 days. The bottom three lines of the output provide the summary statistics that you might include in a manuscript. The adjusted R^2, the F-statistic, the degrees of freedom, and the p-value for the ANOVA. (Just for your information, you should report the adjusted R^2 and not the multiple R^2, as the adjusted R^2 is more conservative and has been *adjusted* for the number of predictors in your model.) As you can gather from the calculations, the Predator treatment had a very strong and significant effect on metamorph SVL, with a p-value of 0.0000224. We can confirm the treatment means with the **summarize()** function. We can even nest that within the **exp()** function to expedite transforming our values back into the real scale of mm.

```
#Nest the mean() function inside the exp() function
#to calculate means on the original scale
temp<-RxP.byTank %>%
  group_by(Pred) %>%
  summarize(mean = exp(mean(log.Age.FromEmergence)))
temp#Look at the summarized data
```

```
## # A tibble: 3 x 2
##   Pred   mean
##   <fct> <dbl>
```

```
## 1 C        34.2
## 2 NL       26.2
## 3 L        15.8
```

Another way to view the summary statistics for the model as a whole is with the **Anova()** function from the *car* package, which gives just the necessary info that you might include in a manuscript: degrees of freedom (both numerator and denominator), F-statistic and p-value.

```
library(car)
Anova(lm1)
```

```
## Anova Table (Type II tests)
##
## Response:  log.Age.FromEmergence
##              Sum Sq Df F value     Pr(>F)
## Pred          9.710  2  12.389 2.244e-05 ***
## Residuals 29.392 75
## ---
## Signif. codes:   0 '***' 0.001 '**' 0.01 '*' 0.05 '.'
    0.1 ' ' 1
```

If you were to report these summary statistics in a paper, you would write it as so: $F_{2,75} = 12.39$, $P = 0.00002$.

We can also visualize these differences, either with a boxplot as before or with a density plot (Figure 5.5).

```
pred.plot1<-qplot(data=RxP.byTank,
                  x=Pred,
                  y=log.Age.FromEmergence,
                  geom="boxplot",
                  fill=Pred)
pred.plot2<-qplot(data=RxP.byTank,
                  x=log.Age.FromEmergence,
                  geom="density",
                  fill=Pred,
                  alpha=0.5)
plot_grid(pred.plot1,pred.plot2, ncol=2)
```

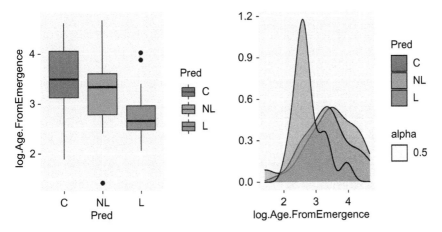

Figure 5.5 In both of these plots, we can clearly see that the predator treatment had a strong effect on froglet age at metamorphosis. In particular, the lethal predator treatment (L) caused froglets to emerge earliest, whereas tadpoles raised with nonlethal (NL) predators emerged next and were intermediate with control (C) animals, who took the longest to leave the water.

5.6.1 Diagnostic plots

It is always a good idea to look at the diagnostic plots for your model. By default, the **plot()** function will give you four diagnostic plots of your model (see Figure 5.6).

1. The residuals plotted against the fitted values (treatment means). This is used to check for consistency of variance across your treatments.

2. A quantile-quantile normal distribution plot (called a Q-Q plot for short), plotting standardized residuals against the theoretical quantiles from a normal distribution. This is used to check for normality of errors. If your model fits well, you expect the real residuals and theoretical residuals to fall on a 1-to-1 line.

3. The square root of the absolute value of the standardized residuals plotted against the fitted values (treatment means). Similar to the first plot, this is used to check for constancy of errors across treatments.

4. "Cook's distance," which is a measure of the influence of each observation on the parameter estimates.

```
plot(lm1)
```

These plots look pretty good (Figure 5.6). The first and third show a relatively even spread of points across our three treatment means. The second plot shows a good similarity between our standardized residuals and the straight line derived from a normal distribution. This indicates good model fit. If the points in the Q-Q plot deviate considerably from the line, your model does not fit a normal distribution well. The last plot does not show us much in this case, but that is good. If there were points with too much leverage, it would be obvious.

One important thing that you probably noticed is that a few of the points have numbers next to them, specifically you can see the numbers 29, 46, and 49 next to three points in each plot. What does this mean? In an attempt to help you figure out if there are any issues with your model, R will highlight the three points (three is the default, but know that you can tell it to request more if you want) with the most extreme (either positive or negative) residual values. The numbers correspond to rows in the data frame, and so this shows us that the data in rows 29, 46, and 49 are the most different from what the model predicts. That does not necessarily indicate a problem, and in this case, it is fair to conclude that these data are not particularly out of sync with everything else.

┤ **Box 5.7 - What is the Q-Q plot?** ├

All of the different diagnostic plots tell you useful information about your model fit, but I find the quantile-quantile (Q-Q) plot the most useful. Understanding what it is plotting will help you see why that is.

Imagine you run a linear regression and obtain a slope and an intercept, which then allows you to plot the **line of best** fit through your scatter of points. The line of best fit is so called because it minimizes the distance between each of your data points and the line. The distance that remains between each of your datapoints and the regression is called a **residual**. Importantly, when taken all together your residuals should be normally distributed. Most should be fairly close to 0, meaning that most points are close to the regression line, and you have fewer points that lie farther away from the line. Now, given this expectation, R can calculate what a *perfect, normal* set of residuals would look like that is the same size as your dataset. If you then plot your real residuals against these perfect hypothetical residuals, you can get a quick visual estimation of the normality of your residuals, aka, how well your model fits. If your real residuals match the theoretical residuals well, you will pretty much see a 1:1 straight line. The more the points deviate from a straight line, the more concern you should have that you are not meeting the assumptions of the model.

It is important to recognize that: 1) the Q-Q plot is not a statistical test of model fit, and 2) it takes a while to get comfortable with how much deviation is acceptable. At the end of the day, you have to figure out for yourself what looks like it fits versus what doesn't fit. We will see plenty of examples of Q-Q plots in this book, but you need to be able to justify to yourself, your colleagues, and potentially your reviewers when a model fits well or does not. When I look at Figure 5.6, I see that nearly all of the points lie very close to the 1:1 line (the dashed line), with the exception of the labeled points 46 and 49. That's pretty good, in my opinion. A little squiggling around the dashed line is certainly to be expected.

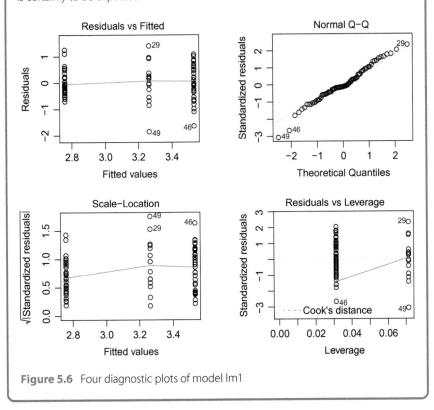

Figure 5.6 Four diagnostic plots of model lm1

5.6.2 A note of caution

Looking back at the summary output for *lm1*, notice that the right side of the "*Coefficients*" section shows several p-values, indicated in the last column which has the heading "$Pr(>|t|)$." It is tempting to look at these and think "wow, my model is really significant!" but you would do best to just ignore these in most cases. The first line, which in the output above shows a t-value of 31.920 and an extremely small p-value, is simply a measure of the difference between the baseline level (here the Control treatment) and zero. The subsequent lines are essentially pairwise t-tests comparing the baseline (C) and the given treatment level (NL or L). However, the p-values reported here are not adjusted for the fact that multiple tests are being conducted, and therefore should not be used as such. Furthermore, it does not tell you all pairwise comparisons (i.e., the comparison of NL and L treatments is missing), which makes it particularly unhelpful. Thus, my advice is to ignore these altogether.

5.7 MULTIPLE COMPARISONS

It is often useful (and necessary) to be able to compare the different levels within a single categorical variable. For example, we probably would like to know if our three predator treatments differ significantly from one another. There are two primary ways to do this. 1) You can use the **glht()** function in the **multcomp** package. **multcomp** is short for "multiple comparisons" and **glht()** stands for "general linear hypothesis test." 2) You can use the **emmeans** package to calculate estimated marginal means from the model and pairwise differences between them. Marginal means are estimated means for your treatment groups that account for other potential covariates or random effects in the model. Thus, they are particularly useful when we have more complex models, but they can be used here as well.

In either case, we will be conducting what is known as Tukey's honestly significant difference (HSD) test, which is a statistical comparison of all pairs of groups. Thus, in our example here with the Pred treatment, we have 3 groups, and so we will calculate the difference between C and L, between C and NL, and between NL and L. Another type of comparison is called a Dunnett's test which is when you compare your experimental treatment groups against just the control. These sorts of post-hoc comparisons are important because they provide you with *adjusted p-values* that correct for the fact that you are conducting multiple comparisons. In a nutshell, any time you conduct more and more statistical comparisons, you are increasing your possibility of finding a "significant" p-value of below 0.05 even though the null hypothesis might really be true. Conducting post-hoc tests in R automatically adjusts your p-values to account for this.

First, let's explore the **glht()** function in the ***multcomp*** package.

```
library(multcomp)
#Run a post-hoc test using glht()
ph1<-glht(lml, linfct=mcp(Pred="Tukey"))
summary(ph1)
```

```
##
##    Simultaneous Tests for General Linear Hypotheses
##
## Multiple Comparisons of Means: Tukey Contrasts
##
##
## Fit: lm(formula = log.Age.FromEmergence ~ Pred,
##          data = RxP.byTank)
##
## Linear Hypotheses:
##               Estimate Std. Error t value Pr(>|t|)
## NL - C == 0   -0.2677     0.2006   -1.335   0.3777
## L  - C == 0   -0.7725     0.1565   -4.936   <0.001 ***
## L  - NL == 0  -0.5048     0.2006   -2.516   0.0362 *
## ---
## Signif. codes:  0 '***' 0.001 '**' 0.01 '*' 0.05 '.'
##     0.1 ' ' 1
## (Adjusted p values reported -- single-step method)
```

Here, we have used **glht()** to conduct a Tukey test (all pairwise comparisons) of the three treatments within our predator treatment. The post-hoc comparisons here allow us to conclude that log.Age.FromEmergence differs significantly for Lethal vs Control treatments, and Lethal vs Non-lethal treatment, but that Control and Nonlethal treatments did not differ. You can see that the reported t-statistics for the comparisons between L and C treatments, or NL and C treatments, are the same as those listed in the **summary()** output of the linear model *lm1*. However, here the p-values are adjusted for the multiple tests being run, and so are more appropriate to use. As mentioned above, you can use **glht()** to conduct a Dunnett's test (comparing treatments against a single control) or any other specific pairwise comparison you desire, but you have to code the specific comparisons yourself (see the help file for **glht()** for advice on how to code those). A handy function (particularly if you have many treatments) is **cld()**, which stands for "compact letter display." It gives you a very simple depiction of which treatments are statistically different from one another.

```
cld(ph1)
```

```
##    C   NL   L
## "b"  "b"  "a"
```

Treatments with the same letter are not different from one another. Thus, in this case, we can see that C and NL are not different (both are "b") and both differ from L (which is "a").

Now, let's see how to use **emmeans()** for post-hoc comparisons. The coding is slightly different from using **glht()** but both techniques produce identical results (at least in this case). The summary output below gives the estimated group means (the "*emmean*" column) along with upper and lower 95 % confidence intervals around the estimated mean. These are not particularly meaningful with a simple model like *lm1*, but will come in handy later on. To actually run the Tukey test, you use the function **pairs()**. One difference worth noting is that when we use the **cld()** function with **emmeans()**, groups that are significantly different from one another are listed with *numbers* instead of letters, like when using **glht()**. Thus, in the output below, you can see the column named ".*group*" and the numbers 1 or 2 below it. Any treatments with the same number are not different from one another.

```
library(emmeans)
#Run a post-hoc test using emmeans()
ph1<-emmeans(lm1, specs="Pred")
summary(ph1)
```

```
##  Pred emmean    SE df lower.CL upper.CL
##  C      3.53 0.111 75     3.31     3.75
##  NL     3.26 0.167 75     2.93     3.60
##  L      2.76 0.111 75     2.54     2.98
##
## Confidence level used: 0.95
```

```
#Use the pairs function to compute the Tukey test
pairs(ph1)
```

```
##  contrast estimate    SE df t.ratio p.value
##  C - NL      0.268 0.201 75   1.335  0.3806
##  C - L       0.773 0.157 75   4.936  <.0001
##  NL - L      0.505 0.201 75   2.516  0.0368
##
## P value adjustment: tukey method for comparing a family of 3 estimates
```

```
#You can also use the cld function with emmeans
cld(ph1)
```

```
##  Pred emmean    SE df lower.CL upper.CL .group
##  L      2.76 0.111 75     2.54     2.98  1
##  NL     3.26 0.167 75     2.93     3.60  2
##  C      3.53 0.111 75     3.31     3.75  2
##
## Confidence level used: 0.95
## P value adjustment: tukey method for comparing a family of 3 estimates
## significance level used: alpha = 0.05
```

⊣ Box 5.8 - Take-home message ⊢

- Data from a "normal," or Gaussian, error distribution are those that have a relatively even spread of values above and below the mean. These data can be any real number, positive or negative.
- You can test your data for normality with a normality test, such as the Shapiro-Wilk Test, but it is often better to use a function like **fitdistr()** to estimate which predefined error distribution your data are most similar to.
- Log-transformation is a common technique to help normalize many biological data.
- There are many old non-parametric statistical tests that can analyze non-normally distributed data, but these have fallen out of favor with the rise of Generalized Linear Models (Chapter 7).
- Student's t-test is designed for analyzing the difference in means of normally distributed data in two categories. A t-test is particularly designed for small sample sizes, and large sample sizes are equivalent to a linear model.
- Linear models are one of the most commonly used form of statistics available. Linear models allow you to analyze a normally distributed response variable with any combination of categorical or continuous predictor variables.
- Many functions exist which are useful for all linear models, such as **summary()**, **Anova()** from the *car* package, and **plot()**. Post-hoc tests can be accomplished using the *emmeans* package.

5.8 ASSIGNMENT!

Here are some things to do on your own, to build on the analysis and plotting skills you have just been working on. Remember you can find example answers and code at https://github.com/jtouchon/Applied-Statistics-with-R.

1. Go back and conduct a series of one-way ANOVAs, analyzing the final Mass of metamorphs by each of the three predictors in the dataset—Hatching age, Predator treatment, and Resource level.
   - Make sure you check the final mass data for normality.
   - Make sure you check the diagnostic plots to see if things look okay or really out of whack (if the model really doesn't fit well, it should be obvious).
   - Use what you learned in Chapter 4 to make a bar graph to go with the analysis of each individual predictor.

- For the analysis of Predator treatment, conduct a post-hoc Tukey test to find out which of the three treatments are different from one another.

2. What would have happened if we had not log-transformed the Age.FromEmergence data? Would we have gotten the same results? What does the Q-Q plot look like?

More Linear Models in R!

6 More Linear Models!

The purpose of this chapter is to expand on what you learned in Chapter 5, introducing you to the multitude of linear models you can do in R. This chapter assumes you are using the **RxP.byTank** dataset that was created in Chapter 5.

This chapter will cover the following topics:

1. Analysis of Variance (ANOVA) with more than one predictor
2. Linear regression
3. Analysis of Covariance (ANCOVA)
4. Plotting regressions
5. Predicting values from a model

As a reminder, we are only concerned only with "normal" data for the moment. By normal data, we are referring to data that fit a normal distribution, with an approximately even distribution of values around the mean. Table 6.1 shows the variety of models that we can conduct under

Applied Statistics with R: A Practical Guide for the Life Sciences. Justin C. Touchon, Oxford University Press (2021). © Justin C. Touchon. DOI: 10.1093/oso/9780198869979.003.0006

the **lm()** framework. Note that all we do is change the number and type of predictor variables to do different types of models.

Table 6.1 All the different flavors of linear models you can do

Data Distribution	Response Variable	Predictor Variable	# Predictors	Test
Normal	continuous	discrete	one	One-way ANOVA
Normal	continuous	discrete	multiple	Multi-way ANOVA
Normal	continuous	continuous	one	Linear regression
Normal	continuous	continuous	multiple	Multiple regression
Normal	continuous	discrete & continuous	multiple	ANCOVA

Analysis of Variance (ANOVA) Analysis of Covariance (ANCOVA)

6.1 GETTING STARTED

The model commonly referred to as a linear model, or **lm()**, is one of the most flexible and useful in all of statistics. We will discuss generalized linear models (GLMs) in Chapter 7, which allow us to analyze non-normally distributed data. Before we get started, let's load all the packages you will need here. This is something I like to do at the head of any script file. Remember that if you don't have one of these packages already, you can download it by using the **Package Installer** menu, or simply by typing **install.packages()** at the prompt and put the name of the package in quotes inside the parentheses. If you use the menu, make sure you click the button to "*include dependencies.*" This is important since most packages rely on other packages to run.

```
library(dplyr)
library(ggplot2)
library(tidyr)
library(car)
library(emmeans)
library(multcomp)
library(cowplot)
```

Now that those are loaded, we can begin.

6.2 MULTI-WAY ANALYSIS OF VARIANCE—ANOVA

The linear model we ran Chapter 5 (remember, it was called *"lm1"*) had a single categorical predictor. This type of model is called a one-way Analysis of Variance (ANOVA). But what about when we have multiple categorical predictors that may interact with one another? This is generally referred to as a multi-way ANOVA—a model with two predictors would be a two-way ANOVA, three predictors would be a three-way ANOVA, and so on. Okay, but what does it mean to "interact?"

Box 6.1 - What is an interaction?

In a simple sense, it means that the effect of one predictor depends on another predictor in the model. With a predator-prey system like the tadpole data we are analyzing, you could imagine that in the absence of predators, having high or low resource levels available might greatly influence prey size, but perhaps the presence of predators causes the prey to all just hide and eat less, so having extra food available doesn't matter. Thus, the effect of resources would depend on (aka *interact with*) the presence or absence of predators.

It should hopefully be clear how this can be extrapolated to any number of study systems. Maybe the drug you administer to your mice affects females differently than it does males. In that case, the presence of the drug would *interact* with animal sex. Or perhaps the fruit flies you study are more active at higher temperatures and so the effect of the ethanol treatment you are exposing them to *interacts* with temperature. These are hypotheticals of course, but hopefully you get the point.

For the RxP data, it is very reasonable to expect that the effects of predators on age at metamorphosis or mass at metamorphosis (or any other response variable) might differ under different resource conditions. To code an interaction between two effects in R you use a colon (:). However, R also provides us with a shortcut for writing both individual effects (e.g., Res or Pred) and the interaction between them (e.g., *Res:Pred*). Using the asterisk (*) tells R to look at both individual and interaction effects between whatever predictors you have provided. Thus, *"Res*Pred"* is the same as writing *"Res+Pred+Res:Pred."* This model is therefore looking at the effects of just resources, of just predators, and then of a possible interaction between the two. Here, we will switch from using Age.FromEmergence (which is what we examined in Chapter 5) to Age.DPO, which is more intuitive since it tells us the age when froglets metamorphosed in terms of when they were laid as eggs.

Let's begin with a two-way ANOVA examining the interacting effects of resources and predators on the log-transformed age at metamorphosis.

```
lm2<-lm(log.Age.DPO~Res*Pred, data=RxP.byTank)
summary(lm2)
```

```
##
## Call:
## lm(formula = log.Age.DPO ~ Res * Pred, data = RxP.byTank)
##
## Residuals:
##       Min       1Q   Median       3Q      Max
## -0.60094 -0.12499  0.00249  0.12355  0.60827
##
## Coefficients:
##                Estimate Std. Error t value Pr(>|t|)
## (Intercept)     4.05433    0.05692  71.228  < 2e-16 ***
## ResLo           0.44028    0.08050   5.470  6.2e-07 ***
## PredNL         -0.05312    0.10318  -0.515   0.6082
## PredL          -0.20223    0.08050  -2.512   0.0142 *
## ResLo:PredNL   -0.10170    0.14591  -0.697   0.4881
## ResLo:PredL    -0.28320    0.11384  -2.488   0.0152 *
## ---
## Signif. codes: 0 '***' 0.001 '**' 0.01 '*' 0.05 '.' 0.1 ' ' 1
##
## Residual standard error: 0.2277 on 72 degrees of freedom
## Multiple R-squared:  0.5231, Adjusted R-squared:    0.49
## F-statistic:  15.8 on 5 and 72 DF,  p-value: 1.779e-10
```

If you look at the **summary()** output, it should look basically the same as what we saw in Chapter 5. However, there is one important thing to notice. The last line of the output contains the F-statistic, degrees of freedom, and P-value for the *model as a whole*. This is different from what we will get from a function like **Anova()**, which gives us those statistics for each of our predictors and the interaction. Thus, we can look at the **summary()** output and see that our model as a whole is highly significant in its ability to explain the variation in "*log.Age.DPO*". However, what you would most likely want to report in a manuscript are the statistics for each individual predictor.

```
#Use the Anova() function from the car package
Anova(lm2)
```

```
## Anova Table (Type II tests)
##
## Response: log.Age.DPO
##            Sum Sq Df F value    Pr(>F)
## Res        1.8241  1 35.1871 9.576e-08 ***
## Pred       1.9446  2 18.7561 2.775e-07 ***
## Res:Pred   0.3254  2  3.1384   0.04934 *
## Residuals  3.7324 72
## ---
## Signif. codes: 0 '***' 0.001 '**' 0.01 '*' 0.05 '.' 0.1 ' ' 1
```

This shows that we have very strong effects of both resource and predator treatments and a significant interaction between the two (albeit not a particularly strong one). In other words, predators alter the age at metamorphosis and so do the different resource levels. However, more importantly our model says that the effect of resources on age at metamorphosis depends on the presence of predators. We could also reframe this in the other direction; the effect of predators on age at metamorphosis depends on the amount of food tadpoles are fed.

6.2.1 Interpreting an output with interactions

Once you have multiple, interacting predictors, the model output becomes increasing difficult to interpret. One of the best things you can do is plot the data in order to visualize the interaction, and we will do that in a moment. First though, let's take a moment to really deconstruct the output from **summary()** and see how to interpret the values we are seeing. Just to review, our "*Res*" treatment has two levels (Low and High) and our "*Pred*" treatment has three levels (Control, Nonlethal and Lethal), giving us 6 total combinations of resources-by-predators (e.g., Low-Control, High-Control, Low-Nonlethal, etc.). There are six lines in the "*Coefficients*" section of the **summary()** output. Just like with a one-way ANOVA, R works alphabetically (or at least it works in the order of how you've set up your factor levels, and the default is of course to be alphabetical). Thus, the baseline listed as the "*(Intercept)*" is the High-Control Resource treatment combo. Since our response variable is log-transformed, our model parameters (aka the estimates) are also on a log scale, so we can use the **exp()** function to exponentiate them and put them back on the real scale of days post-oviposition.

```
#Calculate the real values manually
#This is: (Intercept) aka High-Control
exp(4.05433)
```

```
## [1] 57.64653
```

The subsequent lines all describe changes in the effect sizes in relation to that baseline. Similar to the one-way ANOVA we saw in Chapter 5, the treatment listed in the left most column of the "*Coefficients*" section shows what has changed and *only* what has changed. Thus, in this case where we have two predictors, if only one thing is listed that means the other thing is still the same as the baseline. If the baseline is the High-Control, the second line (listed as "*ResLo*") is the Low-Control treatment combo and the third line (listed as "*PredNL*") is the High-Nonlethal treatment combo. Similarly, the fourth line (listed as "*PredL*") is the High-Lethal treatment combo. The last two lines are the Low-Nonlethal and Low-Lethal treatment combos.

Calculating the effect sizes follows similar logic, but with the addition that any modifications to effect sizes include all lower effects. What does that mean? If the baseline (High-Control) is 4.05433 on the log scale, then the Low-Control is 4.05433+0.44028, just like for the one-way ANOVA example described earlier.

```
#ResLo aka Low-Control
exp(4.05433+0.44028)
```

```
## [1] 89.53324
```

The calculations for High-Nonlethal and High-Lethal treatments can be done in the same manner.

```
#PredNL aka High-Nonlethal
exp(4.05433-0.05312)
```

```
## [1] 54.66425
```

```
#PredL aka High-Lethal
exp(4.05433-0.20223)
```

```
## [1] 47.09185
```

However, when we get to the last two lines of the output (listed as "ResLo:PredNL" and "ResLo:PredL"), the correct effect sizes have to be calculated *including earlier modifications*. Thus, for "ResLo:PredNL," we have the baseline of 4.05433, plus the effect of "ResLo" alone (0.44028) plus the effect of "PredNL" alone (−0.05312) plus the combined effect of "ResLo:PredNL" (−0.10170). Similar logic applies for calculating the 'ResLo:PredL' effect.

```
#ResLo:PredNL aka Low-Nonlethal
exp(4.05433+0.44028-0.05312-0.10170)
```

```
## [1] 76.69143
```

```
#ResLo:PredL aka Low-Lethal
exp(4.05433+0.44028-0.20223-0.28320)
```

```
## [1] 55.10167
```

We can confirm that these treatment estimates are indeed correct by using *dplyr* to calculate the means of each treatment (which are on a log-scale). We can also use **summarize()** to go ahead and exponentiate them for us, which puts everything back in terms of days since the eggs were first laid.

```
#Use dplyr to calculate means and exponentiate them
RxP.byTank %>%
  group_by(Res, Pred) %>%
  summarize(exp.mean.Age.DPO = exp(mean(log.Age.DPO)))
```

```
## # A tibble: 6 x 3
## # Groups:   Res [2]
##   Res   Pred  exp.mean.Age.DPO
##   <fct> <fct>            <dbl>
## 1 Hi    C                 57.6
## 2 Hi    NL                54.7
## 3 Hi    L                 47.1
## 4 Lo    C                 89.5
## 5 Lo    NL                76.7
## 6 Lo    L                 55.1
```

The first version might seem like an unnecessarily cumbersome way to calculate the means of your treatment groups, and it is. However, working from simpler to more complex model structures provides us with a way to really understand the output that R gives us, which will be very useful to you later on.

6.2.2 Posthoc comparisons when you have interactions?

Is it possible to use the **glht()** or **emmeans()** functions described in Chapter 5 to do post-hoc tests and understand which of our six treatment combinations are significantly different from one another? The answer is both yes and no. First, we can use either function to look at the effect of one predictor while averaging across the levels of the other predictor. However, that may not be what we really want to do, especially in an instance like this where we have a significant interaction.

emmeans() allows us to specifically look at one predictor *across* the levels of the other predictor, which is pretty handy. In the following code, we will conduct Tukey's 'honestly significant differences' (HSD) post-hoc test, which compares each level of the factor against all other levels. We specify which predictor we want to look for differences in with the "*specs=*" argument, and we can tell the model to run the Tukey test separately for each level of a separate predictor with the "*by=*" argument.

Now we have a choice to make. Do we analyze the effect of the three predator treatments within the two resource treatments, or do we see what effect resource levels have within each predator treatment? Although in reality you would want to choose one approach or the other, here we can do both and see what it tells us about the data.

First, let's see how predators affected age at metamorphosis within each resource level.

```
#Use the 'by=' argument to do post-hoc tests for one
#factor within each of the levels of the other factor
ph2<-emmeans(lm2, specs="Pred", by="Res")
pairs(ph2)
```

```
## Res = Hi:
##  contrast estimate     SE df t.ratio p.value
##  C - NL     0.0531 0.1032 72 0.515   0.8644
##  C - L      0.2022 0.0805 72 2.512   0.0374
##  NL - L     0.1491 0.1032 72 1.445   0.3234
##
## Res = Lo:
##  contrast estimate     SE df t.ratio p.value
##  C - NL     0.1548 0.1032 72 1.500   0.2968
##  C - L      0.4854 0.0805 72 6.030   <.0001
##  NL - L     0.3306 0.1032 72 3.204   0.0057
##
## P value adjustment: tukey method for comparing a family of 3 estimates
```

This output definitely helps us understand the nature of our interaction. It appears that predators had a greater effect on age at metamorphosis under low resource conditions as compared to high resource conditions. In high resources, only the Control and Lethal predator treatments are different, whereas when tadpoles were fed less the Lethal predators also differed from the Nonlethal, caged predators.

Now, let's see what it tells us if we analyze what effect the two resource levels had within each predator treatment.

```
#Here we have flipped it around and are looking
#at the effect of resources within each predator treatment
ph2<-emmeans(lm2, specs="Res", by="Pred")
pairs(ph2)
```

```
## Pred = C:
##  contrast estimate      SE df t.ratio p.value
##  Hi - Lo    -0.440 0.0805 72 -5.470  <.0001
##
## Pred = NL:
##  contrast estimate      SE df t.ratio p.value
##  Hi - Lo    -0.339 0.1217 72 -2.782  0.0069
##
## Pred = L:
##  contrast estimate      SE df t.ratio p.value
##  Hi - Lo    -0.157 0.0805 72 -1.951  0.0549
```

This is also informative. What we can see is that in the absence of predators, getting more food really mattered and had a strong effect on time to metamorphosis. When tadpoles were exposed to nonlethal, caged predators there was also an effect of resource level, but it wasn't as strong as for the controls. However, when faced with lethal predation the effect of resources was what we might call "marginally significant." No matter how you define it, we can see that lethal predators reduced the impact of resources.

We can visualize the differences amongst treatments by plotting boxplots of the six treatment combinations. Let's plot both ways of thinking about the interaction, as worked out in the previous posthoc analyses. I've also added some extra code to show how you can easily customize the colors in a figure made with *ggplot2*. Specifically, we've added the function **scale_fill_manual()**, which allows us to set unique colors for the levels of whatever we specified the original plot to fill by. In the following first line (the **qplot()** line) we say we want to fill the boxes based on the "*Res*" treatment. Since "*Res*" has two levels (Lo and Hi) we need to provide **scale_fill_manual()** with two values for colors, and those colors have to be concatenated together into a vector with the **c()** function. Recall you can specify colors in many ways in R, using named colors, the **rgb()** function, the hexadecimal coding system, etc. Lastly, notice that I've tacked on the function **theme_cowplot()** which just provides some elegant and clean thematic elements to the plot, such as removing the grid lines and grey background behind the plot. There are many built in themes in ***ggplot2*** which you can read about with a quick Google search (try something like "***ggplot2*** themes"), but I particularly like the ***cowplot*** package (as mentioned many times already).

```
#First, let's plot the effect of Res within Pred treatments
a<-qplot(data=RxP.byTank, x=Pred, y=log.Age.DPO, fill=Res,
      geom="boxplot",
      ylab="log Age at metamorphosis (DPO)")+
  scale_fill_manual(values=c("seagreen",
                             "skyblue"))+
  theme_cowplot()+
  ggtitle(label="Effect of Res within Pred")
#Next, let's plot the effect of Pred within Res levels
b<-qplot(data=RxP.byTank,
         x=Res,
         y=log.Age.DPO,
         fill=Pred,
         geom="boxplot",
         ylab="log Age at metamorphosis (DPO)")+
  scale_fill_manual(values=c("light green",
                             "forest green",
                             "dark green"))+
  theme_cowplot()+
  ggtitle(label="Effect of Pred within Res")
plot_grid(a,b, ncol=2)
```

Both of these plots are particularly illuminating and demonstrate the significant and interacting effects of resources and predators (see Figure 6.1). In the first plot, we see that across all predator treatments, being fed low resources greatly delayed age at metamorphosis, but that this effect was dramatically reduced in the Lethal predator treatment (the difference in Hi and Lo bars is much less). Similarly, in the second plot we can see that

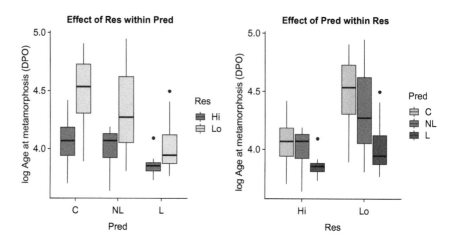

Figure 6.1 The interacting effects of resource levels and predation risk on time to metamorphosis for *A. callidryas* tadpoles. Each of these plots helps demonstrate the interaction between resources and predators

in Hi resources the C and NL treatments are almost identical, whereas in the Lo resource treatment NL is much more intermediate between C and L.

6.2.3 Posthoc comparisons of all six treatments at once

In some cases, you might want to compare all six treatment combos against one another. It doesn't really make sense to me to do that here, but I will show you how to do it in case it is useful to you at some point. In order to do this, we have to make a dummy variable that combines our two treatments ("*Res*" and "*Pred*") into a single variable with six levels, i.e., each of our "*Res*" x "*Pred*" treatment combinations. Let's call the dummy variable "*ResPred*."

One way to do this would be with the **paste()** function, which simply allows us to paste two groups of text together. We can nest the **paste()** command within the function **as.factor()** to ensure that our new dummy variable is a factor. The **paste()** function takes 3 arguments: the two vectors you are pasting together, and some text you will use to separate the text from the vectors. In the following code I've separated the "*Res*" and "*Pred*" values with a hyphen, but you could easily use something else (e.g., a space, a period, an underscore, nothing at all, etc.).

```
RxP.byTank$ResPred<-as.factor(paste(RxP.byTank$Res,
                                    RxP.byTank$Pred, sep="-"))
```

If you want to do this within the ***tidyverse*** there is a function in the package ***tidyr*** (which loads automatically when you load ***tidyverse***) called **unite()** which can accomplish this task. The important thing to know about **unite()** is that you have to provide the name of the column you are creating in quotes, the two or more columns you are uniting together, and if you want to retain the original columns in the data frame (the default is to get rid of them). You can also specify what you want to use to separate the values from the two columns you are uniting but know that the default is an underscore (_).

```
RxP.byTank <- RxP.byTank %>%
              unite(col="ResPred", Pred, Res, sep="-", remove=F)
```

It is important to note that this model is mathematically identical to *lm2*, which was coded with the interaction between resource and predator treatments. Now, we can create a new **lm()** object and run post-hoc tests on the six treatment combinations.

```
lm3<-lm(log.Age.DPO~ResPred, data=RxP.byTank)
ph3<-emmeans(lm3, specs="ResPred")
pairs(ph3)
```

## contrast		estimate	SE	df	t.ratio	p.value
## (C-Hi)	- (C-Lo)	-0.44028	0.0805	72	-5.470	<.0001
## (C-Hi)	- (L-Hi)	0.20223	0.0805	72	2.512	0.1342
## (C-Hi)	- (L-Lo)	0.04515	0.0805	72	0.561	0.9932
## (C-Hi)	- (NL-Hi)	0.05312	0.1032	72	0.515	0.9954
## (C-Hi)	- (NL-Lo)	-0.28547	0.1032	72	-2.767	0.0748
## (C-Lo)	- (L-Hi)	0.64252	0.0805	72	7.982	<.0001

```
## (C-Lo) - (L-Lo)    0.48543 0.0805 72  6.030 <.0001
## (C-Lo) - (NL-Hi)   0.49340 0.1032 72  4.782 0.0001
## (C-Lo) - (NL-Lo)   0.15482 0.1032 72  1.500 0.6651
## (L-Hi) - (L-Lo)   -0.15708 0.0805 72 -1.951 0.3802
## (L-Hi) - (NL-Hi)  -0.14912 0.1032 72 -1.445 0.6995
## (L-Hi) - (NL-Lo)  -0.48770 0.1032 72 -4.727 0.0002
## (L-Lo) - (NL-Hi)   0.00797 0.1032 72  0.077 1.0000
## (L-Lo) - (NL-Lo)  -0.33062 0.1032 72 -3.204 0.0238
## (NL-Hi) - (NL-Lo) -0.33859 0.1217 72 -2.782 0.0720
##
## P value adjustment: tukey method for comparing a family
## of 6 estimates
```

That output gives us the significance of every pair of levels in "*PredRes*", which is a bit overwhelming. This is definitely a time to use the **cld()** function. Recall that groups that are significantly different from one another are listed with *numbers* in **emmeans()**. You want to examine the numbers in the column named ".*group*." Any treatments with the same number are not different from one another.

```
cld(ph3)
```

```
## ResPred emmean     SE df lower.CL upper.CL .group
## L-Hi      3.85 0.0569 72     3.74     3.97  1
## NL-Hi     4.00 0.0861 72     3.83     4.17  12
## L-Lo      4.01 0.0569 72     3.90     4.12  1
## C-Hi      4.05 0.0569 72     3.94     4.17  12
## NL-Lo     4.34 0.0861 72     4.17     4.51   23
## C-Lo      4.49 0.0569 72     4.38     4.61    3
##
## Confidence level used: 0.95
## P value adjustment: tukey method for comparing a family
## of 6 estimates
## significance level used: alpha = 0.05
```

Hopefully the previous output demonstrates why the **cld()** function is so useful, even if this particular model is a little difficult to interpret! We can see that none of the high resource treatments differ from one another, nor do they differ from the Low-Lethal treatment. The Low-Control treatment differs from every other treatment except Low-Nonlethal, which makes sense as they reached metamorphosis way later than any of the other treatments.

6.3 LINEAR REGRESSION

So far, we have discussed scenarios where you have categorical predictors. But, what about when you have a continuous predictor? As long as your response variable is normally distributed, that is a *linear regression*. As mentioned previously, in R a linear regression is just another form of **lm()**. With a linear regression we can ask, for example, if size at metamorphosis (SVL.final) is influenced by the length of the larval period (age.DPO). Once again, we are going to use the log-transformed versions of these variables.

You might be wondering why we are going to use the data on final SVL as our response variable since we saw in Chapter 5 that log-transformation did not make it normal. It improved the normality, but a Shapiro-Wilks test still said it was significantly unlikely that those data came from a normal distribution. There are two reasons to use SVL as our response variable. 1) Biologically speaking, it doesn't make much sense to think of the size at metamorphosis affecting the time it took to get to metamorphosis. Instead, we would likely expect the relationship to go in the other causal direction. 2) Even if our data are not perfectly normal, we can always run a model and evaluate the fit using the diagnostic plots. The **lm()** is extremely robust and if the model fits well, we are in good shape. Let's try, okay?

```
lm4<-lm(log.SVL.final~log.Age.DPO, data=RxP.byTank)
summary(lm4)
```

```
##
## Call:
## lm(formula = log.SVL.final ~ log.Age.DPO, data = RxP.byTank)
##
## Residuals:
```

```
##        Min         1Q    Median        3Q       Max
## -0.182298 -0.049070  0.000516  0.038342  0.159057
##
## Coefficients:
##               Estimate Std. Error t value Pr(>|t|)
## (Intercept)    3.44001    0.09211  37.345  < 2e-16 ***
## log.Age.DPO   -0.11624    0.02232  -5.208 1.58e-06 ***
## ---
## Signif. codes: 0 '***' 0.001 '**' 0.01 '*' 0.05 '.' 0.1 ' ' 1
##
## Residual standard error: 0.06244 on 76 degrees of freedom
## Multiple R-squared:  0.263,   Adjusted R-squared:  0.2533
## F-statistic: 27.12 on 1 and 76 DF,   p-value: 1.581e-06
```

```
Anova(lm4)
```

```
## Anova Table (Type II tests)
##
## Response: log.SVL.final
##              Sum Sq Df F value     Pr(>F)
## log.Age.DPO 0.10576  1  27.123 1.581e-06 ***
## Residuals   0.29633 76
## ---
## Signif. codes: 0 '***' 0.001 '**' 0.01 '*' 0.05 '.' 0.1 ' ' 1
```

Indeed, "*log.Age.DPO*" has a highly significant effect on "*log.SVL.final*." We can also see in the Coefficients section of the **summary()** output that the regression has a negative slope, indicating that as age increases SVL decreases (more on this is to follow). As mentioned previously, we should defintely look at the diagnostic plots to evaluate how well the model fits.

```
plot(lm4)
```

Overall, this looks quite good. As discussed in Chapter 5, R will label observations that may be problematic. Looking across the first plots (Figure 6.2), we can see that the observations in rows 26, 49, and 75 have been

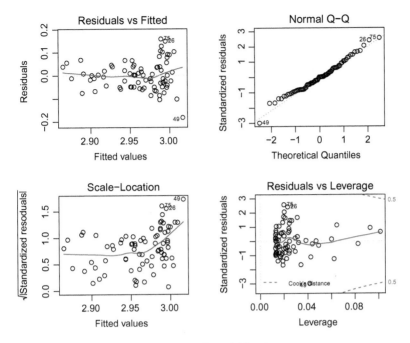

Figure 6.2 Four diagnostic plots of model **lm4**

flagged. The location on the x-axis (called "*fitted values*") tells us where the model expects those points to fall according to the regression, and the residuals tells us how far above or below that expectation those points are. We can see that row 49 is much smaller than expected, whereas rows 75 and 26 are larger than expected. Are these problematic? These points actually look pretty good in the Q-Q plot and do not look like a big deal according to the leverage plot (the fourth panel), so it's somewhat hard to say. To really know if these observations are indeed influencing our model, we should re-run the model while excluding them and see if it changes our results in any appreciable way.

Box 6.2 - Removing potentially problematic data?

What is the easiest way to exclude a few data points? As we've seen in previous chapters, we could use square brackets ("[ ]"s) or **filter()** to subset out the rows we do not want to include in our dataset. However, whereas we previously used indexing to create a new data frame in the memory, here we can use indexing to exclude the potentially problematic data directly in the model we are running. By changing the "*data=*" argument in our model specification from "*data=RxP.byTank*" to "data=RxP.byTank[-c(26,49,75),]," we can run the model but without rows 26, 49, and 75. What we are doing with that code is we are using a minus sign (-) to remove whatever is in the vector (what is inside the **c()** function). This is the same idea as subsetting we did before, just with a slightly different technique. One reason to do this is that we do not have to permanently modify our dataset or create a new object in the memory. We just run the model, but say to exclude the three rows we want to get rid of.

```
lm4.1<-lm(log.SVL.final~log.Age.DPO,
          data=RxP.byTank[-c(26,49,75),])
summary(lm4.1)
```

```
##
## Call:
## lm(formula = log.SVL.final ~ log.Age.DPO,
##     data = RxP.byTank[-c(26, 49, 75), ])
##
## Residuals:
##       Min        1Q    Median        3Q       Max
## -0.102504 -0.047131  0.001842  0.038959  0.129485
##
## Coefficients:
##               Estimate Std. Error t value Pr(>|t|)
## (Intercept)    3.44064    0.08238   41.77  < 2e-16 ***
## log.Age.DPO   -0.11681    0.01990   -5.87 1.19e-07 ***
## ---
## Signif. codes:  0 '***' 0.001 '**' 0.01 '*' 0.05 '.' 0.1 ' ' 1
##
## Residual standard error: 0.05431 on 73 degrees of freedom
## Multiple R-squared:  0.3207, Adjusted R-squared:  0.3114
## F-statistic: 34.46 on 1 and 73 DF,  p-value: 1.19e-07
```

```
Anova(lm4.1)
```

```
## Anova Table (Type II tests)
##
## Response: log.SVL.final
##              Sum Sq Df F value   Pr(>F)
## log.Age.DPO 0.10164  1  34.457 1.19e-07 ***
## Residuals   0.21533 73
## ---
## Signif. codes:  0 '***' 0.001 '**' 0.01 '*' 0.05 '.' 0.1 ' ' 1
```

Removing those three observations does not have a dramatic effect on our model. Without them, our R-squared is admittedly about 6% greater, and our F-statistic is larger, but the slope is still in the same direction and the fundamental relationship being described has not changed. Thus, it is not a problem to leave the three observations in the model.

Before moving on, I'll introduce one more function that is useful for simple data manipulation like we've just done. In the **dplyr** package there is a function called **slice()** which allows you to subset your data based on row number, just like we've done with square brackets. The advantage of using **slice()** is that it is a little more intuitive to read. **slice()** is essentially a special case of **filter()**, which we saw in Chapter 3. For example, to remove the observations in rows 26, 49, and 75, we could do the following:

```
temp <- RxP.byTank %>% slice(-c(26,49,75))
lm4.1<-lm(log.SVL.final~log.Age.DPO, data=temp)
Anova(lm4.1)
```

```
## Anova Table (Type II tests)
##
## Response: log.SVL.final
##                Sum Sq Df F value   Pr(>F)
## log.Age.DPO 0.10164  1  34.457 1.19e-07 ***
## Residuals   0.21533 73
## ---
## Signif. codes: 0 '***' 0.001 '**' 0.01 '*' 0.05 '.' 0.1 ' ' 1
```

6.3.1 Interpreting the regression

Since we have determined our original model **lm4** is not problematic, we can proceed with interpreting the output. Recall that the formula for a straight line is $y = slope * x + intercept$ (I learned it as $y=mx+b$ back in high school, but it is the same equation). When you look at the summary output, you can see it looks exactly the same as what we saw when we had categorical predictors (i.e., the one-way and multi-way ANOVA's described earlier and in Chapter 5). However, now our "*(Intercept)*" is the actual intercept for the regression line, and the parameter listed for "*log.Age.DPO*" is our slope. The summary output tells us that for every unit increase in the log of Age.DPO, we have a unit decrease in the log of SVL.final. Another way to write this would be as follows:

log.SVL.final = 3.44001 - 0.11624*log.Age.DPO.

Since this is a log-log model (i.e., both the predictor and response variables were log-transformed), we know that the relationship between Age and SVL is fundamentally nonlinear. We can plot this in two ways. The easiest thing to do is to keep the axes and values on a log-scale, which allows us to plot the regression as a straight line. However, without *too much* difficulty we can plot the regression back on the values of the original data (days and mm, respectively), which will make the model more visually intuitive.

6.3.2 Plotting the regression

The function **abline()** plots a straight line on an existing scatterplot. **abline()** needs two arguments, an intercept (a) and a slope (b). Thus, in R's terminology, the equation for a straight line is $y = a + bx$ (Figure 6.3). You can also pass other optional plotting parameters if you want to change the line width, style, color, and so on.

```
plot(log.SVL.final~log.Age.DPO, data=RxP.byTank)
abline(3.44001,-0.11624)
```

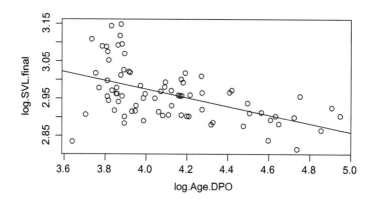

Figure 6.3 A simple regression plot made in base graphics. Aren't base graphics the worst?

A shortcut for when you have a simple linear regression (i.e., you only have a single continuous predictor) is to just put the name of the model object in the parentheses (e.g., *abline(lm4)*). The function knows to get the intercept and regression from the model. This does not work with models with multiple predictors.

In order to plot the regression back on the original data, we have to do some work ourselves. More importantly, this is a really great way to explore how to interpret a linear regression. Recall that the equation for the regression listed previously is "*log.SVL.final = 3.44001 - 0.11624* log.Age.DPO*." If we wanted to know the predicted log of the final size of a metamorph emerging at, for example, 40 days post-oviposition, we can easily do the following:

```
3.4401 - 0.11624 * log(40)
```

```
## [1] 3.011305
```

That answer is the log SVL, so we can put that back into real units (mm) by exponentiating it.

```
exp(3.011305)
```

```
## [1] 20.31389
```

Thus, the predicted size of a metamorph emerging 40 days after oviposition is 20.3 mm. The same logic applies for any time *days post-oviposition* (DPO) we might choose. For example, to calculate the predicted size of metamorphs emerging at 60, 80, or 100 days post-oviposition, we can easily type the following:

```
exp(3.4401 - 0.11624 * log(60))
```

```
## [1] 19.37868
```

```
exp(3.4401 - 0.11624 * log(80))
```

```
. ## [1] 18.74137
```

```
exp(3.4401 - 0.11624 * log(100))
```

```
## [1] 18.2615
```

You can see that as DPO increases, the predicted SVL of a metamorph decreases. Now let's take this a step further. Instead of merely passing R a single value for DPO, we can pass it a vector. For example, to predict the sizes of metamorphs for every day between 40 and 60 DPO, we can do the following:

```
exp(3.4401 - 0.11624 * log(40:60))
```

```
## [1] 20.31389 20.25566 20.19900 20.14383 20.09007 20.03766
## [7] 19.98653 19.93663 19.88790 19.84029 19.79375 19.74824
## [13] 19.70372 19.66014 19.61747 19.57567 19.53471 19.49457
## [19] 19.45519 19.41657 19.37868
```

The colon (:) in the previous example is a shortcut that R uses to make a vector of whole integers between two values (go ahead and type any two numbers with a colon between them into the console and see what happens. What happens if the first number you type is larger than the second one?). Let's make a vector of those predicted values for every day between 35 and 145 DPO, which is essentially the length of our x-axis. Once we have that vector, we can use the function **lines()** to plot the vector on top of our scatterplot. **lines()** is a function that requires only two things: a vector of values for x-coordinates and a matching vector of y-values (they must be the same length). Since we used "*35:145*" to define the range of predicted SVL's, that will also be the range of our x-coordinates (Figure 6.4).

```
#plot raw values
plot(SVL.final~Age.DPO, data=RxP.byTank)
#make a vector of the predicted SVL's
SVL.line<-exp(3.44001-0.11624*log(35:145))
#Use the predicted values to plot a line
lines(x=35:145, y=SVL.line, lwd=2)
```

How would we go about making a plot like this in **_ggplot2_**? To simply add a straight linear regression is very easy, but it is less straightforward to add the regression from the model on the original values. In order to make a simple linear regression, simply add the extra function **geom_smooth()** after your plot, along with a statistical "*method*" for determining what kind of line to plot. The default is to make a Lowess curve, which might not be what you want. Here we will add "*method='lm'.*" Essentially, we are telling the plot to run a linear model inside the plotting function and draw the resulting linear regression. This works with both **qplot()** and **ggplot()** (see Figure 6.5). Also note that by default **geom_smooth()** will plot a confidence interval around the line. To get rid of that, you could add the argument "*se=F*" within **geom_smoooth()**, if you want.

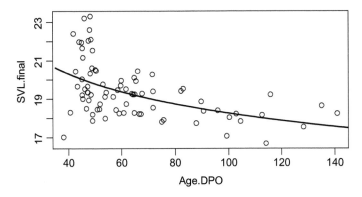

Figure 6.4 A scatterplot of SVL against age DPO, with a regression line plotted from **lm4**. Here, the figure is plotted using the original values. Doesn't this figure look nicer than Figure 6.3? It is important to remember that they are technically showing us the exact same thing.

```
qplot(data=RxP.byTank,
      x=log.Age.DPO,
      y=log.SVL.final,
      geom="point")+
  geom_smooth(method="lm")+
  theme_cowplot()
```

That was so easy! What about putting it back on the raw values? That's trickier. Most importantly, we have to make a two-column data frame that contains the x- and y-coordinates of the line, just like we did before. If we wanted to add the confidence interval, we would have to calculate the upper and lower bounds of the confidence interval ourselves and add it to the plot as a "ribbon" with the function **geom_ribbon()** (see Figure 6.6).

```
#First, make a data frame of the x- and y-coordinates
SVL.line<-data.frame(X = 35:145,
                     Y = exp(3.44001-0.11624*log(35:145)))
ggplot()+ #Leave out the data for now
  geom_point(data = RxP.byTank,
             aes(x = Age.DPO,
                 y = SVL.final))+ #points
  geom_line(data = SVL.line,
            aes(x = X,
                y = Y))+ #the line
```

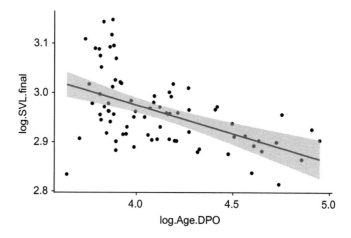

Figure 6.5 It's so easy to add a nice looking regression to a plot using ***ggplot2***!

```
ylab("Final metamorph SVL (mm)")+ #y-axis labels
xlab("Age at metamorphosis (DPO)")+ #x-axis labels
theme_cowplot() #make it look nice
```

6.4 ANALYSIS OF COVARIANCE (ANCOVA)

What happens when we have a combination of continuous and categorical predictors? This sort of model is called an *analysis of covariance* or *ANCOVA* for short. The name comes from the fact that it is an ANOVA that also contains a *covariate*, or a continuous variable. For example, perhaps we would like to know if the relationship we have been examining between age DPO and SVL differs when tadpoles are raised in different predator environments? When you code the predictors in an ANCOVA, you want to remember to code the covariate before any categorical variables. Note that now it is even more important to use the **Anova()** function from the *car* package to calculate summary statistics, as using **anova()** would give you misleading summary stats.

```
#Now let's make an ANCOVA that looks at the
#effects of categorical and continuous data together
lm5<-lm(log.SVL.final~log.Age.DPO*Pred, data=RxP.byTank)
summary(lm5)
```

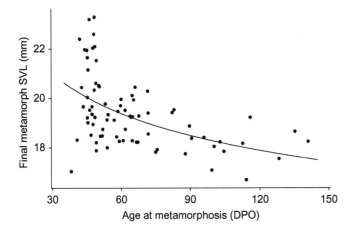

Figure 6.6 It's less easy to add a line that is not directly plotted from a function internally in **geom_smooth()**. But it can be done without *too much difficulty*!

```
##
## Call:
## lm(formula = log.SVL.final ~ log.Age.DPO * Pred,
##    data = RxP.byTank)
##
## Residuals:
##      Min        1Q    Median        3Q       Max
## -0.122214 -0.029839  0.002357  0.036826  0.122903
##
## Coefficients:
##                      Estimate Std. Error t value Pr(>|t|)
## (Intercept)           3.08467    0.12387  24.902   <2e-16 ***
## log.Age.DPO          -0.03974    0.02890  -1.375   0.1733
## PredNL               -0.05606    0.21020  -0.267   0.7905
## PredL                 0.67183    0.24458   2.747   0.0076 **
## log.Age.DPO:PredNL    0.02000    0.04982   0.401   0.6893
## log.Age.DPO:PredL    -0.14883    0.06090  -2.444   0.0170 *
## ---
## Signif. codes: 0 '***' 0.001 '**' 0.01 '*' 0.05 '.' 0.1 ' ' 1
##
## Residual standard error: 0.05227 on 72 degrees of freedom
## Multiple R-squared:  0.5107, Adjusted R-squared:  0.4767
## F-statistic: 15.03 on 5 and 72 DF,  p-value: 4.324e-10
```

For the sake of comparison, here is the output from both the **anova()** and **Anova()** functions.

```
anova(lm5)
```

```
## Analysis of Variance Table
##
## Response: log.SVL.final
##                  Df  Sum Sq  Mean Sq F value    Pr(>F)
## log.Age.DPO       1 0.105757 0.105757 38.7050 2.908e-08 ***
## Pred              2 0.079871 0.039935 14.6155 4.706e-06 ***
## log.Age.DPO:Pred  2 0.019731 0.009865  3.6105   0.03204 *
## Residuals        72 0.196732 0.002732
```

```
## ---
## Signif. codes: 0 '***' 0.001 '**' 0.01 '*' 0.05 '.' 0.1 ' ' 1
```

```
Anova(lm5)
```

```
## Anova Table (Type II tests)
##
## Response: log.SVL.final
##                   Sum Sq Df F value    Pr(>F)
## log.Age.DPO      0.019898  1  7.2824   0.00867 **
## Pred             0.079871  2 14.6155 4.706e-06 ***
## log.Age.DPO:Pred 0.019731  2  3.6105   0.03204 *
## Residuals        0.196732 72
## ---
## Signif. codes: 0 '***' 0.001 '**' 0.01 '*' 0.05 '.' 0.1 ' ' 1
```

Notice that the significance of "*log.Age.DPO*" is much higher in the **anova()** output than the **Anova()** version. This is because the **anova()** version calculates significance of each predictor in a step-by-step manner, so the significance of the first predictor (in this case "*log.Age.DPO*") is considered the only thing in the model. The **Anova()** output, on the other hand, calculates significance assuming everything else in the model, which is why the latter predictors have the same stats in both summary outputs. In general, the approach used by **Anova()** is more conservative and therefore preferrable.

6.4.1 Interpreting an ANCOVA

Now that we know which summarized output to use, what does it mean? How do you interpret the fact that both predictors and their interaction are significant in an ANCOVA? Essentially, the significance of the first predictor ("*log.Age.DPO*") means that if you pooled together all the meta-morphs from the three predator treatments, then there is a significant relationship between age and size. Similarly, the significant effect of *Pred* basically means that if you collapse the variation across ages and just focus on the predator effect, then the mean SVL at metamorphosis varies across the three treatments. Lastly, the significant interaction is saying that *the relationship between metamorph SVL and the length of the larval period differs depending on the presence of lethal, nonlethal, or no predators in the environment.* Another way to phrase this is that the slopes of the three regressions (one for each of our three predator treatments) are different. We don't really know which of the three are different at this point, simply *that there is a difference amongst the three treatments.* Whereas we used the function **emmeans()** to conduct post-hoc analyses of an ANOVA, now we can use the function **emtrends()** (conveniently also found in the ***emmeans*** package) to test for which slopes are different from which others.

6.4.2 Post-hoc anlyses of an ANCOVA

We can still use **emmeans()** to conduct Tukey post-hoc comparisons as before, but now the function will average across the effect of "*log.Age.DPO*," which is probably not that useful. The function even gives you a warning to let you know that the interaction means you probably should not be putting too much stock in whatever results you are seeing.

```
emmeans(lm5, specs="Pred")
```

```
## NOTE: Results may be misleading due to involvement in
## interactions

##  Pred emmean     SE df lower.CL upper.CL
##  C      2.92 0.0103 72     2.90     2.94
##  NL     2.95 0.0142 72     2.92     2.98
##  L      2.98 0.0135 72     2.95     3.01
##
## Confidence level used: 0.95
```

In order to test for differences in the slopes of our three treatments, we need to add an argument to the function **emtrends()**. The code is same as for **emmeans()**, but now we say that we want the slopes of the predictor specified by the "*var=*" argument to be compared *across* the levels of the predictor specified by "*specs=*." We can use the function **pairs()** as above to conduct the Tukey test. If we wanted, we could use the **cld()** function as above.

```
#Use emtrends() to test for differences in slopes
ph4<-emtrends(lm5, specs="Pred", var="log.Age.DPO")
pairs(ph4)
```

```
##  contrast estimate    SE df t.ratio p.value
##  C - NL    -0.020 0.0498 72  -0.401  0.9151
##  C - L      0.149 0.0609 72   2.444  0.0443
##  NL - L     0.169 0.0672 72   2.511  0.0376
##
## P value adjustment: tukey method for comparing a family
## of 3 estimates
```

This post-hoc analysis tells us that the slopes (i.e., the relationship between age and size) of the control and nonlethal treatments do not differ, but that both differ from that of the lethal predator treatment. Let's plot these relationships in order to visualize them!

6.4.3 Interpreting the summary output for an ANCOVA

Recall earlier how we calculated the parameter estimates for a two-way ANOVA. The **summary()** output gives us the coefficients for each of our predictor variables in relation to a baseline, and the baseline is the alphabetically first treatment level (which in our case is C, the control). The same logic applies here, but now we have three regressions, each corresponding to our three predator treatments.

The first two lines of the **summary()** output labeled "*(Intercept)*" and "*log.Age.DPO*" correspond to the intercept and slope for our baseline regression, the Control. Anything subsequently labeled with the covariate, "*log.Age.DPO*," will be a slope and anything that just lists the name of a Predator treatment will be an intercept. Thus, for example, the line labeled "*PredL*" is the modified intercept for the Lethal Predator treatment, and the line labeled "*log.Age.DPO:PredL*" is the corresponding modified slope. Importantly, this allows us to actually calculate the predicted size of any metamorph at any date and to see how those sizes change based on predator treatment. We could use **abline()** to plot the three regressions on the log-log scales as before, but moving forward we will just use *ggplot2*, as it is much better for this sort of thing. Earlier, we used **geom_smooth()** to add the regression to the plot. Since the *ggplot2* functions are very smart, if we facet our plot or we use "*col=*" to make multiple plots, **geom_smooth()** will inherit those instructions and make multiple regressions (Figure 6.7)! It's great, it's amazing, it's brilliant!

```
qplot(data=RxP.byTank,
      x=log.Age.DPO,
      y=log.SVL.final,
      geom="point",
      col=Pred) +
   geom_smooth(method="lm", se=F) +
   theme_cowplot()
```

Notice that by default, the regression lines are only plotted to the extent of the data points for each predator treatment, and not to for the full extent of the data. If you add the argument "*fullrange=T*" to the **geom_smooth()** function, you will plot lines for the full range of data.

Box 6.3 - Interpreting the plot of an ANCOVA

So, what does Figure 6.7 tell us? In essence, what we can glean from the figure is that the presence of Lethal predators fundamentally changes the relationship between larval period and size at metamorphosis. Nonlethal predators slightly increase the size of metamorphs, but the way that size changes when tadpoles are in the water longer isn't really affected. Lethal predators, on the other hand, cause a drastic increase in SVL and a general shortening of the larval period and this results in a fundamentally different relationship between size and larval period.

6.5 THE **predict()** FUNCTION

Now that we have seen how to calculate our predicted values by hand, and how to translate those into plotted curves, let's calculate them using a function built in to R: **predict()**. The most useful reason to use **predict()** is that it provides an easy way to calculate 95% confidence intervals (CIs)

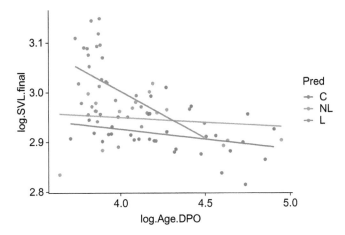

Figure 6.7 A scatterplot of metamorph SVL against age DPO, with regression lines representing the ANCOVA in **lm5**. We can see that this confirms the results from the **emtrends()** function. The slopes of the C and NL treatments are not very different from one another, whereas both are very different than that of the L treatment.

around regression lines. It is also very handy for non-linear functions (which we will explore more in Chapter 7). As we've seen, it is easy to use **geom_smooth()** to plot regressions with confidence intervals. However, sometimes you might actually want to know what those numbers are, which you can't get from the plot (although of course you can visually estimate them).

Using the **predict()** function takes a little getting used to, but it is very useful once you get the hang of it. **predict()** essentially requires just two arguments, but in practical use you will most likely want to provide it with three arguments. You need to provide **predict()** with 1) the model you are predicting values from, 2) a set of "new data" to assign the new values to, and 3) whether or not you want to calculate confidence intervals (the default is to not provide them). In Chapter 7, we will add a fourth argument, which is the type of output to provide.

The most confusing aspect of this is the "new data" argument. If you look back at our model lm4, we have one predictor, the Age.DPO predictor (on the log scale of course). The "new data" is essentially what we did earlier when we created the "*SVL.line*" object, where we provided a length of x-axis values for the regression equation to use to calculate the y-axis values of final SVL. Now we can let **predict()** do that math for us. Let's create a new data frame with hypothetical values for our x-axis (the log of age at metamorphosis post-oviposition), and a blank column for **predict()** to fill in for the corresponding final SVL. First, let's figure out the smallest and largest values of our x-axis values using the **range()** function. You could also use the functions **min()** and **max()** if you wanted.

```
range(RxP.byTank$log.Age.DPO)
```

```
## [1] 3.640506 4.948068
```

Okay, our smallest value is around 3.64 and our largest is about 4.95. To make a sequence of numbers that spans that range, we can use the function **seq()**, which allows you to specify a minimum, a maximum, and either how many numbers you want it to output or the interval between the numbers you want. Thus, if we want to make a sequence between 3.64 and 4.95 every 0.01 values, we can type:

```
seq(from=3.64, to=4.95, by=0.01)
```

```
##   [1] 3.64 3.65 3.66 3.67 3.68 3.69 3.70 3.71 3.72 3.73
##  [11] 3.74 3.75 3.76 3.77 3.78 3.79 3.80 3.81 3.82 3.83
##  [21] 3.84 3.85 3.86 3.87 3.88 3.89 3.90 3.91 3.92 3.93
##  [31] 3.94 3.95 3.96 3.97 3.98 3.99 4.00 4.01 4.02 4.03
##  [41] 4.04 4.05 4.06 4.07 4.08 4.09 4.10 4.11 4.12 4.13
##  [51] 4.14 4.15 4.16 4.17 4.18 4.19 4.20 4.21 4.22 4.23
##  [61] 4.24 4.25 4.26 4.27 4.28 4.29 4.30 4.31 4.32 4.33
##  [71] 4.34 4.35 4.36 4.37 4.38 4.39 4.40 4.41 4.42 4.43
##  [81] 4.44 4.45 4.46 4.47 4.48 4.49 4.50 4.51 4.52 4.53
##  [91] 4.54 4.55 4.56 4.57 4.58 4.59 4.60 4.61 4.62 4.63
## [101] 4.64 4.65 4.66 4.67 4.68 4.69 4.70 4.71 4.72 4.73
## [111] 4.74 4.75 4.76 4.77 4.78 4.79 4.80 4.81 4.82 4.83
```

```
## [121]  4.84 4.85 4.86 4.87 4.88 4.89 4.90 4.91 4.92 4.93
## [131]  4.94 4.95
```

Okay, now we are ready to make our data frame. We should name the new variables in our data frame *exactly what they are called in the model* so that **predict()** will know what they are. We can use the function **head()** to look at the first 6 lines and verify that it worked.

```
lm4.newdata<-data.frame("log.Age.DPO" = seq(from=3.64,
                                             to=4.95,
                                             by=0.01))
head(lm4.newdata)
```

```
##   log.Age.DPO
## 1        3.64
## 2        3.65
## 3        3.66
## 4        3.67
## 5        3.68
## 6        3.69
```

Now for the last step, use **predict()** to calculate our regression curve and associated confidence intervals. To calculate 95% CI's, add the argument "*se.fit=TRUE.*" Note that the default is a 95% CI, but you can change it to be whatever you would like.

```
lm4.predicted<-predict(lm4,
                       newdata=lm4.newdata,
                       se.fit=T)
str(lm4.predicted)
```

```
## List of 4
##  $ fit          : Named num [1:132] 3.02 3.02 3.01 3.01 ...
##   ..- attr(*, "names")= chr [1:132] "1" "2" "3" "4" ...
##  $ se.fit       : Named num [1:132] 0.0127 0.0126 0.0124 ...
##   ..- attr(*, "names")= chr [1:132] "1" "2" "3" "4" ...
##  $ df           : int 76
##  $ residual.scale: num 0.0624
```

If you look at the structure of the output from **predict()** you see that it is a list. We haven't dealt much with lists yet, as they can be kind of a pain to work with. The good thing about them though is that, as opposed to a data frame, they can hold multiple objects of different lengths. The first item *"fit"* is the estimated mean values themselves and the second item *"se.fit"* is the ± CI around those estimated means.

We can of course create more complex blank data frames to feed into **predict()**, for more complex functions. For example, let's calculate the estimated size ± CI of metamorphs emerging between 35 and 145 days post-oviposition from each of the three predator treatments.

In the next section, I have used the function **expand.grid()**, which is a cool function that creates *every combination* of whatever you give it. By providing it a vector of our 3 predator treatments, and a vector of the logs of each whole number between 35 and 145, the function creates a data frame with all the combinations. You will probably note that this is different than what we did earlier in the chapter, where we actually input the logged versions of the smallest and largest values. I thought you might want to see another way to tackle the same problem, so here you go. You can use the functions **head()** or **tail()** to verify that it worked.

```
#First create a data frame to provide values to predict()
lm5.newdata<-expand.grid("log.Age.DPO"=log(35:145),
                         Pred=c("C","NL","L"))
head(lm5.newdata)
```

```
##    log.Age.DPO Pred
## 1     3.555348    C
## 2     3.583519    C
## 3     3.610918    C
## 4     3.637586    C
## 5     3.663562    C
## 6     3.688879    C
```

```
tail(lm5.newdata)
```

```
##        log.Age.DPO Pred
## 328      4.941642    L
## 329      4.948760    L
## 330      4.955827    L
## 331      4.962845    L
## 332      4.969813    L
## 333      4.976734    L
```

```
#Now we create a data frame that houses the predicted values
#and matches the variables in our data frame lm5
lm5.predicted<-predict(lm5, newdata=lm5.newdata, se.fit=T)
str(lm5.predicted)
```

```
## List of 4
## $ fit           : Named num [1:333] 2.94 2.94 2.94 2.94 ...
##   ..- attr(*, "names")= chr [1:333] "1" "2" "3" "4" ...
## $ se.fit        : Named num [1:333] 0.0227 0.022 0.0213 ...
##   ..- attr(*, "names")= chr [1:333] "1" "2" "3" "4" ...
## $ df            : int 72
## $ residual.scale: num 0.0523
```

Once again, the output from **predict()**, in this case an object we've called "*lm5.predicted*," is a list. After getting the predicted values, we can make new columns in the lm5.newdata data frame in order to make it easy to plot this all. In the next example, we've exponentiated everything to put it back on the real scale so it is easier to interpret, as well as calculated the upper and lower bounds of the confidence intervals ("*fit*" + "*se.fit*" and "*fit*" - "*se.fit*," respectively).

```
#First, exponentiate the fitted values
lm5.newdata$predict.SVL<-exp(lm5.predicted$fit)
#Next, calculate the upper and lower bounds of the CIs
lm5.newdata$predict.SVL.CIupper<-exp(lm5.predicted$fit+
                              lm5.predicted$se.fit)
lm5.newdata$predict.SVL.CIlower<-exp(lm5.predicted$fit-
                              lm5.predicted$se.fit)
#Lastly, exponentiate the age data so it is back
#on the original scale
lm5.newdata$Age.DPO<-exp(lm5.newdata$log.Age.DPO)
str(lm5.newdata)
```

```
## 'data.frame':    333 obs. of  6 variables:
##  $ log.Age.DPO        : num   3.56 3.58 3.61 3.64 3.66 ...
##  $ Pred               : Factor w/ 3 levels "C","NL","L": 1 1 .
##  $ predict.SVL        : num   19 19 18.9 18.9 18.9 ...
##  $ predict.SVL.CIupper: num   19.4 19.4 19.3 19.3 19.3 ...
##  $ predict.SVL.CIlower: num   18.6 18.5 18.5 18.5 18.5 ...
##  $ Age.DPO            : num   35 36 37 38 39 40 41 42 43 44 ...
```

6.6 PLOTTING WITH **ggplot()** INSTEAD OF **qplot()**

Now, we can plot these regressions with their confidence intervals. The following code introduces *a lot* of new aspects of plotting with ***ggplot2***, so it is worthwhile to walk through it all a bit.

1. First and foremost, we've switched from using the function **qplot()** to **ggplot()**. **ggplot()** offers a lot more flexibility and, while generally similar, has some important differences compared to **qplot()**. The most important difference is that you don't specify the geom in the first function anymore, you specify a separate geom function that you "add" to the original function. We've actually already seen this, when we added **geom_smooth()** to our plot earlier. So, if you don't specify the geom, what do you do in the first **ggplot()** function? Essentially, you specify what's going to be used to build the figure. You assign the data, and you assign the aesthetics, which is the ***ggplot2*** way of saying you use the **aes()** argument/function to set up which will be the data used for the x- and y-axes and if you are going to be coloring or filling your data by any other variables. Then, you add to it various functions in order to make the plot look like what you want. An example of this was used to make Figure 6.5, but here we will walk through the code much more slowly.

2. A second thing to know is that, like base graphics, the order of objects does somewhat matter. For example, in the following code I've plotted the ribbon first, then the points, then the regression

line. That essentially layers the shaded confidence interval *behind* the points, which are *behind* the regression line. Some orders don't matter though. You can change the colors assigned to a fill anywhere and it will affect everything that has come before or after it.

3. A third important point is that as opposed to making figures in base graphics where everything is fixed once it is plotted, when you make a graphic with ***ggplot2*** the things you add on to the original function will adjust how everything looks. It all gets interpreted together. Thus, for example, when you add a legend it adjusts everything so that the plot fits without the legend being on top of anything else.

Okay, so what does the following code do?

- The first line sets up the basics of the plot: the x-axis is "*Age.DPO*," the y-axis is "*SVL.final*," and we want to color points based on "*Pred*," which will create three groups of points, and these data come from the "*RxP.byTank*" data frame.
- The next function is **geom_ribbon()**, which is how we plot confidence intervals. Now, since these data *do not* come from RxP.byTank we have to specify a new data object and specify that we do not want to inherent the aesthetics from the original call (using the argument "*inherit.aes=F*"). We can then assign new aesthetics to define the ribbon, which is the max and min on the y-axis, as well as that we want to fill the ribbons based on "*Pred*." We've also specified that we want our CI's to be 40% transparent (with the "*alpha=0.4*" argument).
- Next, we have said that we want to plot points, with the **geom_point()** function. Since there is nothing in the parentheses, everything necessary to plot the points will be inherited from the first line, the **ggplot()** function.
- Next, we've said that we want to plot a line with the **geom_line()** function. Just like the CI's, the data for the line do not come from

the originally specified data frame, so we again have to say we don't want to inherit the aesthetics and then we have to go on to specify our x and y variables and that we want to color our lines by the "*Pred*" treatment.

- Hopefully by now you've noticed that we've *filled* some things and *colored* other things. Some geoms are filled, like the ribbon, and some are colored, like lines or points. Thus, the next two lines specify manually assigned colors for our three treatments with the "*values=*" arguments. This is done with **scale_fill_manual()** and **scale_color_manual()**, respectively. The functions are basically the same, except they work for filled or colored objects (imagine that!). Each one will automatically create a legend, which is great, except for when you have to use both and don't want two of the same legend! We certainly don't need that. Hence, in the **scale_fill_manual()** function we've specified "*guide=F*" in order to suppress the creation of the legend. More importantly, you can see in the **scale_color_manual()** function we are able to see the "*labels*" for our legend, so that instead of just the values from our data (C, NL, and L) we can enter in more meaningful names. We can also assign a more meaningful name to the legend than the default. Note that while in this code I chose to create a legend with useful names in the **scale_color_manual()** function, we easily could have done the same thing with **scale_fill_manual()**.
- We end by creating new labels for our x-axis and y-axis and setting the theme to **theme_cowplot()**, which is just a simple plot style without grid lines or anything.

There is a lot here! A great way to explore and understand code like this is to delete parts bit by bit to see what they do, or more generally just change the code to see what happens. What happens if you move the **geom_ribbon()** line to the end? What happens if you delete the "*fill=*" parts anywhere? What happens if you don't tell **geom_line()** to not inherit the aesthetics from the original function? All of this code takes a little while

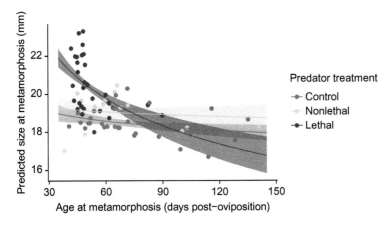

Figure 6.8 The predicted size of metamorphosing *A. callidryas* across the entire larval period. The presence of Lethal predators has a very strong effect on animals metamorphosing during the first month or so, but this effect largely disappears in animals that stay in the water longer.

to get used to, but you can make really great figures once you learn it (see Figure 6.8).

```
ggplot(data=RxP.byTank, aes(x=Age.DPO,
                            y=SVL.final,
                            col=Pred))+
  geom_ribbon(data=lm5.newdata,
              inherit.aes=F,
              aes(x=Age.DPO,
                  ymax=predict.SVL.CIupper,
                  ymin=predict.SVL.CIlower,
                  fill=Pred),
              alpha=0.4)+
  geom_point()+
  geom_line(data=lm5.newdata,
            inherit.aes=F,
            aes(x=Age.DPO,
                y=predict.SVL,
                col=Pred))+
  scale_fill_manual(values=c("seagreen",
                             "skyblue",
                             "dark red"),
                    guide=F)+
  scale_color_manual(values=c("seagreen",
                              "skyblue",
```

```
                               "dark red"),
                 labels=c("Control",
                          "Nonlethal",
                          "Lethal"),
                 name="Predator treatment")+
  ylab("Predicted size at metamorphosis (mm)")+
  xlab("Age at metamorphosis (days post-oviposition)")+
  theme_cowplot()
```

Now, let's back up a moment and reconsider why it is useful to know how to use the **predict()** function. There are three reasons (at least in my mind).

1. From a statistical perspective, we often talk about *significance*, right? Simply put, if the 95% confidence intervals of two means do not overlap, then you can pretty safely assume that those two groups are *significantly different*. Thus, being able to calculate those CIs and plot them is a useful tool for being able to estimate statistically significant groups.

2. If you want to plot a confidence interval around a curve that is not calculated directly from the model, such as we did earlier, you will have to calculate the CI yourself. When you have multiple curves, such as in Figure 6.8, doing this is much easier with **predict()**.

3. In many fields, it is useful to be able to forecast (aka *predict*) something beyond just the data you have in hand. With a function like **predict()**, we can *extrapolate* what might happen to our response variable even in conditions we may not have actually measured. What might happen if the temperature in your treatment was cooler or warmer? What about if you increase the drug dosage you're administering to your mice? Those are the sorts of things you can help address with the **predict()** function.

Box 6.4 - Take-home message

- The linear model, **lm()**, is one of the most important and universally used models in modern statistical analysis. This single framework gives you the option to conduct what we often refer to as one-way or more ANOVA, ANCOVA, and linear regression, all with a unified and simple coding structure.
- Posthoc tests for linear models are a breeze with **emmeans()** and **emtrends()**.

- Understanding significant interactions can be somewhat tricky. Your goal should be to identify what is driving the statistical nature of the interaction, as well as to interpret what it means for your data. In addition to conducting posthoc analyses, one of the most revealing things you can do is simply to plot the data, and interpret the interaction visually.
- Make sure to use **Anova()** instead of **anova()** in order to get the most conservative and accurate estimates of significance for the predictors in your models.
- Plotting non-linear regression lines on to your raw data is a little more difficult than you might hope, but it isn't that hard. Use functions like **predict()** to calculate the curves and confidence intervals and use **ggplot()** to plot them beautifully.

6.7 ASSIGNMENT!

Here are some exercises to try on your own, to build on the data analysis and plotting skills you have just worked on. Complete the following tasks and answer the associated questions for yourself, then look at the Github page for this book (https://github.com/jtouchon/Applied-Statistics-with-R) for sample code and answers.

1. At the end of the Chapter 5, you analyzed the log of final mass by each of the three categorical predictors independently. What happens if you include all three predictors in the model at the same time? Does the significance of any of the individual predictors change when you include more variables? If so, what does that tell you about how your interpretation of one variable might change when you include other variables in the model?

2. Combine your knowledge of **predict()** and **ggplot()** to plot an ANCOVA different than that represented by **lm5**. Make sure it is plotted on the original scale of whatever variables you choose and that it has confidence intervals. Make sure to add interesting colors and useful axes, etc.

3. Explore the statistical analyses shown here but with different interactions. Pick two different questions to ask. For example, does Hatching age interact with Age.DPO to influence the size of the tail remaining when froglets crawl out of the water? How is age at

metamorphosis influenced by predators and resources, or predators, resources, and hatching age? These are just examples, there
are lots of questions you can ask! Just make sure that you are clear
about what question you are asking and then conduct the analysis
to answer it.

Generalized Linear Models (GLM)

7 Generalized Linear Models (GLM)

The purpose of this chapter is to begin introducing you to some more advanced statistical analyses using R. It is assumed you are using the **RxP**, **RxP.clean**, and **RxP.byTank** datasets created in the earlier chapters.

7.1 UNDERSTANDING NON-NORMAL DATA

In Chapters 5 and 6, we discussed the utility of the **lm()** function for analyzing normally distributed data. Normal data are, as you hopefully remember, data where you have a relatively even spread of values above and below the mean. These values can be any real number, positive or negative, and most values are close to the mean with the probability of obtaining values decreasing as you get farther away from the mean. However, in the real world, data are often not normally distributed. In these cases, we can turn to *Generalized Linear Models* (GLMs), which extend the modeling framework of **lm()** to many other error structures ("error" refers to the way the data are spread around the mean). The basic function for running a GLM is the aptly named **glm()** function.

Applied Statistics with R: A Practical Guide for the Life Sciences. Justin C. Touchon, Oxford University Press (2021). © Justin C. Touchon. DOI: 10.1093/oso/9780198869979.003.0007

Box 7.1 - Making non-normal data normal

GLMs work via a "link" function, which transforms the data to a normal scale. For example, with a binomial GLM (also called a logistic regression) we are not modeling the data (usually a 0 or 1) but are instead modeling the log odds of an event happening (getting a 1) or not happening (getting a 0). In this case, the log odds of the event occurring is normally distributed. The built in error families available with **glm()** are as follows (although there are others):

- binomial(link = "logit")
- gaussian(link = "identity")
- poisson(link = "log")
- Gamma(link = "inverse")
- inverse.gaussian(link = "1/mu^2")
- quasi(link = "identity", variance = "constant")
- quasibinomial(link = "logit")
- quasipoisson(link = "log")

The link functions written in parentheses next to the previous error families are the defaults for each error family and are automatically assumed, so you do not need to actually write them in your model specification. Several families have multiple possible link functions, some of which may be preferred in certain circumstances, so it is good to know the different possible options. That said, I have never needed to use a link function different than the default, so you probably won't either.

Box 7.2 - Understanding error families

So, what do these different error families mean? What do the data look like and how do you know which error function to use? Here are some basic rules of thumb for the most commonly used error distributions (including one that is very useful and not on the previous list).

- Binomial: These are generally data that fall into two categories. For example, animals lived or died, or eggs have hatched or have not hatched, frogs are males or are females. R is pretty smart though and you can also use a binomial error distribution to model proportional data, so long as the proportions are bounded at 0 and 1. We will discuss an example of this later in the chapter.
- Gaussian: This is just a fancy stats way of saying "normal distribution," it is named for Johann Carl Friedrich Gauss. Thus, you can use the **glm()** function to run the exact same linear models we made in Chapters 5 and 6, if you want.
- Poisson: Named for the French mathematician Siméon Denis Poisson, these data are counts of observations which tend to have a tail out to the right. Since they are generally counts of things, they cannot be negative and can only be whole integers.

One of the defining features of a Poisson distribution is that the variance of the data is approximately the same as the mean. Also, R will let you use a Poisson distribution with non-integer values, but it will give you a warning message.

- Negative binomial: This might sound like it is related to the binomial error distribution (and it is), but it is easier to think of this as an "overdispersed Poisson." If a Poisson dataset is one where the mean and variance are the same, a negative binomial dataset is one where the variance is considerably larger than the mean. In other words, the tail of data out to the right is much longer. There are more extreme values. Like a Poisson, the data cannot be negative and should only be integers (although R will once again let you use non-integer data, even though it will yell at you for doing so). Note that the negative binomial distribution has its own special function, **glm.nb()**.
- Gamma: A Gamma distribution is similar in shape to a negative binomial, but is continuous data (i.e., not just integers).
- Lastly, ignore the last three distributions mentioned in Box 7.1 (quasi, quasibinomial, and quasipoisson). I've included them here because you would see them if you look at the help file for the **glm()** function, but no one uses them anymore. The "quasi" models were shortcuts for modeling overdispersed data that were implemented in the 2000's but more advanced and statistically appropriate techniques have been developed, rendering the "quasi" models moot. You can also ignore the inverse gaussian distribution for the most part, at least in my experience.

There are also several other specific functions that have been developed for certain error structures not incorporated in the original **glm()** function, such as the function **betabin()** (in the **aod** package) for analyzing beta-binomial data or the function **betareg()** (in the **betareg** package) for using a beta distribution of proportional data.

7.2 GLMS

Let's create two new variables which are the number of tadpoles that survived to metamorphosis, and the flip side of that, the number that died before metamorphosis. We can once again use the *dplyr* package to accomplish this. However, now instead of using **summarize()** to calculate the **mean()** of each group of data, we can use the function **length()** to calculate how many of each group we have. In the next section, we have calculated the length (i.e., the number of values in a vector) of the summarized "*Ind*" variable. We could use any of the variables there, since **group_by()** essentially subsets the data by each tank and since we are taking the length of the vectors of how many tadpoles are in each tank at metamorphosis, any of the variables should give us the same answer.

First, let's load all the packages we'll need.

```
library(dplyr)
library(ggplot2)
library(car)
library(MASS)
```

```
#Create a new variable and store it as an object called temp
temp <- RxP.clean %>%
        group_by(Tank.Unique) %>%
        summarize(N.alive = length(Ind))
temp
```

```
## # A tibble: 78 x 2
##     Tank.Unique N.alive
##           <int>   <int>
## 1             1      47
## 2             2      42
## 3             3      45
## 4             4      26
## 5             5      40
## 6             6       8
## 7             7      43
## 8             8      39
## 9             9      23
## 10           10      44
## # ... with 68 more rows
```

"*Tank.Unique*" is the individual number of each tank and "*N.alive*" is the total number of metamorphs from each tank. Now that we have these numbers calculated, we want to add this column to our existing RxP.byTank data frame. There are two ways to do this. First, if we can assume that our rows in the two objects are in the *exact same order*, we could just create a new column in RxP.byTank.

```
#When you call a column that does not exist,
#R will just make it
RxP.byTank$N.alive<-temp$N.alive
```

Alternatively, we could modify our code from Chapter 5 when we made the "*RxP.byTank*" data frame to include the command to make the new "*N.alive*" variable. Since you should still have this code in a script somewhere, it is super simple to modify it.

```
RxP.byTank<-RxP.clean %>%
  group_by(Tank.Unique, Pred, Res, Hatch, Block) %>%
  summarize(Age.DPO = mean(Age.DPO),
          Age.FromEmergence = mean(Age.FromEmergence),
          SVL.initial = mean(SVL.initial),
          Tail.initial = mean(Tail.initial),
          SVL.final = mean(SVL.final),
          Mass.final = mean(Mass.final),
          Resorb.days = mean(Resorb.days),
          N.alive = length(Ind))##This line is  new!!
#Remember to reorder the Pred factor
RxP.byTank$Pred<-factor(RxP.byTank$Pred, levels=c("C","NL","L"))
```

It might also be useful to calculate the number of tadpoles that died or were eaten before metamorphosis. Since we know that we started with 50 tadpoles per tank, this is very simple to calculate. In the following code, I've introduced a new function called **glimpse()**, which is essentially the *dplyr* version of **str()**. Have you noticed that if you use **str()** on a tibble, you often get a bunch of gobbledygook at the bottom? I find that mess annoying, so **glimpse()** offers a cleaner way to look at your data frame when it is a tibble (although I will admit I do not like its appearance as much as **str()**).

```
RxP.byTank$N.dead<-50-RxP.byTank$N.alive
glimpse(RxP.byTank)
```

```
## Rows: 78
## Columns: 14
## Groups: Tank.Unique, Pred, Res, Hatch [78]
## $ Tank.Unique     <int> 1, 2, 3, 4, 5, 6, 7, 8, 9, 10, 11...
## $ Pred            <fct> NL, C, C, L, NL, L, NL, C, L, C, ...
## $ Res             <fct> Hi, Hi, Hi, Lo, Hi, Hi, Lo, Lo, H...
```

```
## $ Hatch            <fct> L, E, L, L, E, E, L, E, L, L, E, ...
## $ Block            <int> 1, 1, 1, 1, 1, 1, 1, 1, 1, 1, 1, ...
## $ Age.DPO          <dbl> 47.19149, 45.38095, 53.82222, 56....
## $ Age.FromEmergence <dbl> 13.19149, 11.38095, 19.82222, 22....
## $ SVL.initial      <dbl> 19.42553, 18.40476, 18.92667, 18....
## $ Tail.initial     <dbl> 4.834043, 5.369048, 4.802222, 4.6...
## $ SVL.final        <dbl> 19.65957, 19.00952, 19.12000, 19....
## $ Mass.final       <dbl> 0.4178723, 0.3821429, 0.4117778, ...
## $ Resorb.days      <dbl> 3.489362, 3.785714, 3.511111, 3.6...
## $ N.alive          <int> 47, 42, 45, 26, 40, 8, 43, 39, 23...
## $ N.dead           <dbl> 3, 8, 5, 24, 10, 42, 7, 11, 27, 6...
```

Now we have two new variables which represent the number of red-eyed treefrog tadpoles in each tank that survived to metamorphosis or died before getting there. Let's look at the distribution of the mortality data (see Figure 7.1). Note that in the following histogram I have specified how many bins the data should be placed in for plotting.

```
qplot(data=RxP.byTank,
      x=N.dead,
      geom="histogram",
      bins=10)
```

This shows us that most of the values are fairly small (i.e., low mortality) but that some tanks had very high mortality. There is a long tail of data out

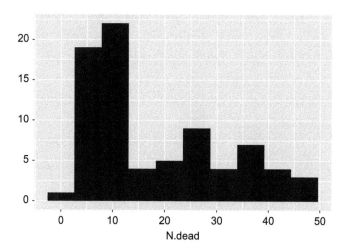

Figure 7.1 The distribution of the number of *A. callidryas* tadpoles that died prior to metamorphosis.

to the right. These data are counts data (whole integers), cannot be negative and fit the general shape of what we expect Poisson or negative binomial data to look like.

How do we decide what error distribution fits best? Recall from an earlier chapter that the function **fitdistr()** from the *MASS* package allows us to evaluate the fit of different error distributions to a given set of data. We can use that here to see which error distribution works best for these data. Note that you have to have a little bit of preliminary knowledge to know which distributions you might want to choose from (don't worry, that knowledge will come in time). Here, let's evaluate four distributions: 1) normal, 2) lognormal, 3) Poisson, and 4) negative binomial. Recall that you can see all of the possible distributions in the help file for **fitdistr()**.

```
#Create 4 objects for evaluating the best
#error distribution for the N.dead variable
fit1<-fitdistr(RxP.byTank$N.dead, "normal")
fit2<-fitdistr(RxP.byTank$N.dead, "lognormal")
fit3<-fitdistr(RxP.byTank$N.dead, "Poisson")
fit4<-fitdistr(RxP.byTank$N.dead, "negative binomial")
#Use AIC() to compare the fit to each distribution
AIC(fit1,fit2,fit3,fit4)
```

```
##       df       AIC
## fit1   2   625.4562
## fit2   2   598.1027
## fit3   1  1060.4417
## fit4   2   598.3479
```

Looking at the Akaike information criteria (AIC) scores, you can see that the Poisson is by far the worst fit (recall that smaller AIC scores are better) and that the lognormal and negative binomial are pretty similar.

⊣ Box 7.3 - Coding with glm() ⊢

Coding a generalized linear model **glm()** is almost identical to coding a linear model with **lm()**. The only real difference is that you include an extra argument *"family="* to define the error distribution family you are using. Aside from that, everything is the same.

Let's explore the differences between the four error distributions described previously. First, let's make four models, each using one of the error distributions, and let's use "*Res*" and "*Pred*" (and their interaction) as our predictor variables. Notice that all models use the following **glm()** function except for the negative binomial distribution which uses a special version of **glm()** that is just for the negative binomial, **glm.nb()**, which is found in the ***MASS*** package. Since the function specifies that it is for a negative binomial, you do not need to specify a family argument, as in the **glm()** function. Also note that we could have used **lm()** for the first two models, since the Gaussian family is the same as a normal distribution.

```
glm.n<-glm(N.dead~Res*Pred, family="gaussian",
        data=RxP.byTank)
glm.ln<-glm(log(N.dead)~Res*Pred, family="gaussian",
        data=RxP.byTank)
glm.p<-glm(N.dead~Res*Pred, family="poisson",
        data=RxP.byTank)
glm.negb<-glm.nb(N.dead~Res*Pred, data=RxP.byTank)
```

7.3 UNDERSTANDING AND INTERPRETING THE GLM

So, you just coded your first GLMs. Hooray! The code for **glm()** *is exactly the same* as **lm()**, except now we have added the extra argument to specify the error family. So simple! In a moment we will examine the **summary()** outputs and you will see that they are remarkably similar to what you have looked at before for linear models.

Now that we've made our models, let's examine their diagnostic plots. Remember, these are all looking at the same data—the number of tadpoles that died before metamorphosis—it's just that they make different assumptions about how the data should be modeled. It's not as simple as comparing two means or something.

```
###Note that this line allows you to
###plot all 4 diagnostic plots at once!
par(mfrow=c(2,2))
plot(glm.n)
```

```
plot(glm.ln)
```

```
plot(glm.p)
```

```
plot(glm.negb)
```

7.3.1 Comparing the diagnostic plots and making a decision

Okay, so what should you be looking at here? As described in an earlier chapter, you are looking for funky patterns and points that are labeled as not fitting particularly well. I usually pay the most attention to the Q-Q plot, which is 2nd plot of the 4 (the upper right panel in each of Figures 7.2–7.5). Recall that in this plot you are looking for the points to fall more or less close to the dashed line. The normal distribution is surprisingly not terrible, given how non-normal the data clearly are. The lognormal is

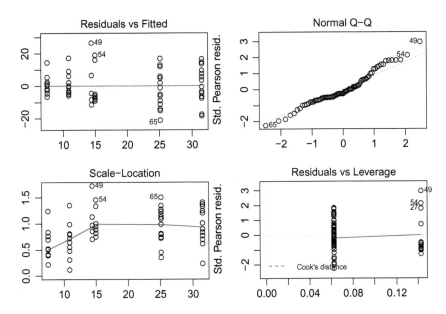

Figure 7.2 Diagnostic plots of a model analyzing the number of tadpoles that died before metamorphosis assuming an underlying **normal distribution** to the data.

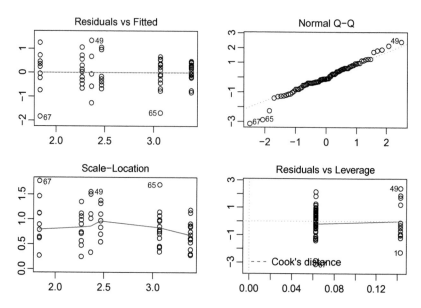

Figure 7.3 Diagnostic plots of a model analyzing the number of tadpoles that died before metamorphosis assuming an underlying **lognormal distribution** to the data.

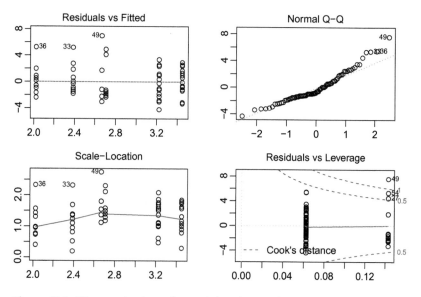

Figure 7.4 Diagnostic plots of a model analyzing the number of tadpoles that died before metamorphosis assuming an underlying **Poisson distribution** to the data.

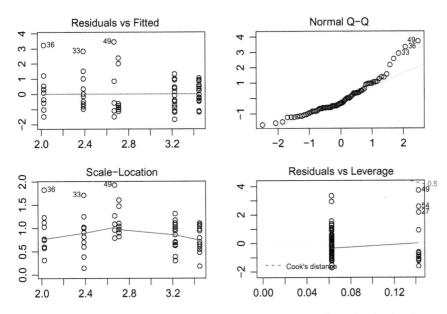

Figure 7.5 Diagnostic plots of a model analyzing the number of tadpoles that died before metamorphosis assuming an underlying **negative binomial distribution** to the data.

really good overall but has a few points that are very far from the line, as do the Poisson and negative binomial. Overall, I would say that the negative binomial and the lognormal appear to fit the best. While the Q-Q plot for the Poisson looks pretty okay, if you look at the Residuals vs Leverage plot (the 4th plot in each set) you can see that there are a number of problematic points in the Poisson model. The dashed red lines indicate that those points are having considerable sway, or leverage, on the model. In general, the negative binomial and lognormal look like the best of these four models.

Now, let's compare the summary outputs for the Poisson and negative binomial models. First, let's walk through the Poisson output.

```
summary(glm.p)
```

```
##
## Call:
## glm(formula = N.dead ~ Res * Pred, family = "poisson",
##     data = RxP.byTank)
##
## Deviance Residuals:
##     Min       1Q    Median        3Q       Max
## -5.2488   -1.3954   -0.7289    1.0861    5.7004
##
## Coefficients:
##               Estimate Std. Error z value Pr(>|z|)
## (Intercept)    2.02320    0.09091  22.255  < 2e-16 ***
## ResLo          0.36326    0.11837   3.069  0.00215 **
## PredNL         0.64601    0.13478   4.793 1.64e-06 ***
## PredL          1.20066    0.10369  11.579  < 2e-16 ***
## ResLo:PredNL  -0.32442    0.18286  -1.774  0.07603 .
## ResLo:PredL   -0.13714    0.13595  -1.009  0.31310
## ---
## Signif. codes: 0 '***' 0.001 '**' 0.01 '*' 0.05 '.' 0.1 ' ' 1
##
## (Dispersion parameter for poisson family taken to be 1)
##
##     Null deviance: 710.99  on 77  degrees of freedom
## Residual deviance: 352.95  on 72  degrees of freedom
## AIC: 712.4
##
## Number of Fisher Scoring iterations: 5
```

First, notice that your output looks almost exactly the same as for the linear models we have looked at before. However, in the world of GLMs we no longer talk about the *variance* of the data in our model, but instead the *deviance* of the data. They mean the same thing. Here, the Null deviance gives us a unitless measure of how much variation there is in the data, and Residual deviance gives us an estimate of how much of the variation is not explained by our model (aka, it is variation that is left over after your model has explained what it can explain).

┤ **Box 7.4 - Understanding overdispersion** ├

As a general rule of thumb, you should hope that your *Residual deviance* is not more than twice your degrees of freedom. In our Poisson model, the residual deviance is nearly 5X our degrees of freedom, which is not good. This is a phenomenon called *overdispersion*, which is when your data has more variation than is expected by the model. How can your data have too much variation you might ask? Since we are modeling a specific error distribution, there are built in assumptions about what those data will look like. For example, a Poisson distribution assumes your mean and variance are roughly the same (it is quite literally the *definition* of a Poisson distribution). If we examine the mean and variance of our N.dead variable, we can see that this assumption is certainly violated.

```
mean(RxP.byTank$N.dead)
```

```
## [1] 18.03846
```

```
var(RxP.byTank$N.dead)
```

```
## [1] 171.1284
```

Luckily for us, smart statisticians have come up with other error distributions to handle such a situation, and an overdispersed Poisson is modeled by a negative binomial error distribution. The flipside of overdispersion is, as you might guess, underdispersion. This is less common and is when your data have substantially less variation than expected for the given error distribution.

Now, let's examine the summary output of the negative binomial model.

```
summary(glm.negb)
```

```
##
## Call:
## glm.nb(formula = N.dead ~ Res * Pred, data = RxP.byTank,
##      init.theta = 4.618042907, link = log)
##
## Deviance Residuals:
##     Min        1Q     Median        3Q        Max
```

```
## -2.5788   -0.7213   -0.3361    0.4428    2.4394
##
## Coefficients:
##                 Estimate Std. Error z value Pr(>|z|)
## (Intercept)       2.0232     0.1476  13.703  < 2e-16 ***
## ResLo             0.3633     0.2027   1.792  0.07308 .
## PredNL            0.6460     0.2503   2.581  0.00984 **
## PredL             1.2007     0.1945   6.174 6.66e-10 ***
## ResLo:PredNL     -0.3244     0.3498  -0.927  0.35371
## ResLo:PredL      -0.1371     0.2695  -0.509  0.61081
## ---
## Signif. codes: 0 '***' 0.001 '**' 0.01 '*' 0.05 '.' 0.1 ' ' 1
##
## (Dispersion parameter for Negative Binomial(4.618) family
##  taken to be 1)
##
##     Null deviance: 156.263  on 77  degrees of freedom
## Residual deviance:  80.091  on 72  degrees of freedom
## AIC: 554.85
##
## Number of Fisher Scoring iterations: 1
##
##
##              Theta:  4.618
##          Std. Err.:  0.962
##
##  2 x log-likelihood:  -540.854
```

You can see that the Residual deviance is much lower than it was in the Poisson model, meaning that the model has done a better job of accounting for the variation in the model. We will proceed with this model for the examples here.

The output from a GLM is on the scale of the link function. The link for a negative binomial is the same as for Poisson, a log-link. Thus, you would use the **exp()** function to transform the parameters back onto the original scale of the data (the numbers of tadpoles that died). Note that calculating the various effect sizes is exactly the same as what we did for **lm()** in Chapters 5 and 6. Recall that to calculate the estimates from the summary output, interaction effects are the sum of all the lower-level effects. For example:

```
exp(2.0232)#High control
```

```
## [1] 7.562486
```

```
exp(2.0232 + 0.3633)#Low control
```

```
## [1] 10.87536
```

```
exp(2.0232 + 0.6460)#High nonlethal
```

```
## [1] 14.42842
```

```
exp(2.0232 + 1.2007)#High lethal
```

```
## [1] 25.12592
```

```
exp(2.0232 + 0.6460 + 0.3633 -0.3244)#Low nonlethal
```

```
## [1] 15.00075
```

```
exp(2.0232 + 0.6460 + 0.3633 -0.1371)#Low lethal
```

```
## [1] 18.09074
```

7.4 CALCULATING STATISTICAL SIGNIFICANCE WITH GLMS

Just as with linear models, we can use the **Anova()** function in the *car* package to calculate the significant predictors in our model. You can use the **anova()** but I highly recommend the version with the capital "A".

```
Anova(glm.negb)
```

```
## Analysis of Deviance Table (Type II tests)
##
## Response: N.dead
##             LR Chisq Df Pr(>Chisq)
## Res            3.979  1    0.04607 *
## Pred          71.617  2  2.809e-16 ***
## Res:Pred       0.873  2    0.64615
## ---
## Signif. codes: 0 '***' 0.001 '**' 0.01 '*' 0.05 '.' 0.1 ' ' 1
```

As you can see, both predators and resources (although to a lesser extent) affected the number of tadpoles that survived to metamorphosis. The interaction between predators and resources was not significant, indicating that the effect of resource level on survival is consistent across the different predator treatments. We can see this if we plot the data (see Figure 7.6).

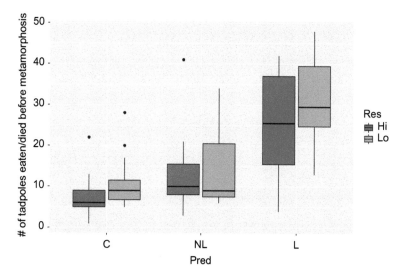

Figure 7.6 Predators, and to a lesser extent resources, greatly affect how many tadpoles died prior to metamorphosis. Tadpoles with higher levels of resources had greater survival (i.e., fewer were eaten), while lethal predators consumed LOTS of tadpoles.

```
qplot(data=RxP.byTank,
      x=Pred,
      y=N.dead,
      ylab="# of tadpoles eaten/died before metamorphosis",
      geom="boxplot",
      fill=Res)
```

Box 7.5 - The ramifications of choosing the right error distribution

It is worthwhile to take a moment to examine what would be the effect of choosing the different error distributions previously illustrated. If we just look at the output from the **Anova()** functions, we can see the effect of assuming different error distributions.

Anova(glm.n)

```
## Analysis of Deviance Table (Type II tests)
##
## Response: N.dead
##            LR Chisq Df Pr(>Chisq)
## Res           3.462  1     0.0628 .
## Pred         64.316  2  1.082e-14 ***
## Res:Pred      0.961  2     0.6186
## ---
## Signif. codes:  0 '***' 0.001 '**' 0.01 '*' 0.05 '.' 0.1 ' ' 1
```

Anova(glm.ln)

```
## Analysis of Deviance Table (Type II tests)
##
## Response: log(N.dead)
##            LR Chisq Df Pr(>Chisq)
## Res           5.696  1     0.0170 *
## Pred         62.342  2  2.901e-14 ***
## Res:Pred      0.721  2     0.6972
## ---
## Signif. codes:  0 '***' 0.001 '**' 0.01 '*' 0.05 '.' 0.1 ' ' 1
```

Anova(glm.p)

```
## Analysis of Deviance Table (Type II tests)
##
## Response: N.dead
##            LR Chisq Df Pr(>Chisq)
```

```
## Res           18.01  1  2.202e-05 ***
## Pred         336.88  2  < 2.2e-16 ***
## Res:Pred       3.15  2     0.2067
## ---
## Signif. codes:  0 '***' 0.001 '**' 0.01 '*' 0.05 '.' 0.1 ' ' 1
```

```
Anova(glm.negb)
```

```
## Analysis of Deviance Table (Type II tests)
##
## Response: N.dead
##          LR Chisq Df Pr(>Chisq)
## Res         3.979  1    0.04607 *
## Pred       71.617  2  2.809e-16 ***
## Res:Pred    0.873  2    0.64615
## ---
## Signif. codes:  0 '***' 0.001 '**' 0.01 '*' 0.05 '.' 0.1 ' ' 1
```

What we can see in the previous outputs is that choosing the wrong error distributions can impact your view of which variables significantly affect the response variable. If we had just ignored the error distribution and gone with a standard two-way Analysis of Variance (ANOVA) (the "*glm.n*" model) we would have concluded that there was not a significant effect of resources. On the other hand, had we chosen to go with a Poisson model (the "*glm.p*" model) we would have concluded that resources had an extremely significant effect. This is one of the effects of overdispersion: it can dramatically inflate your estimation of significance. If we had not examined the summary output from the Poisson model carefully and seen that the residual deviance was so large, we might not have realized the model was not appropriate. Based on examining the Q-Q plot and based on our understanding of the shape of the data, we should use the negative binomial or lognormal models ("*glm.negb*" or "*glm.ln*"), both of which conclude that resources have a significant, albeit relatively small, effect. Hopefully this example demonstrates the importance of being thorough and critical when you examine your data and choose the way you will analyze them.

7.5 CODING THE DATA AS A BINOMIAL GLM

This is perhaps a bit of a side note, but we could also think about analyzing these data with a binomial GLM (also called a Logistic Regression) since they are proportional data. Traditionally, binomial data are 0s and 1s. However, R allows you to model proportional data using a binomial error family. In this case, the data are coded as a two-column table featuring wins and losses, or animals that lived and died. Thus, the two columns add up to the total starting number of individuals. This is important; you cannot code the data as individuals that died and total starting number (well, you could code it that way, but your results would be wrong). To make the two-column table, you bind together two columns with the **cbind()** function.

```
glm.b<-glm(cbind(N.alive,N.dead)~Res*Pred, family="binomial",
          data=RxP.byTank)
summary(glm.b)
```

```
##
## Call:
## glm(formula = cbind(N.alive, N.dead) ~ Res * Pred,
##       family = "binomial", data = RxP.byTank)
##
## Deviance Residuals:
##     Min      1Q   Median      3Q      Max
## -7.804  -1.666    1.018   1.788    6.470
##
## Coefficients:
##                 Estimate Std. Error z value Pr(>|z|)
## (Intercept)      1.72483    0.09868  17.480  < 2e-16 ***
## ResLo           -0.44454    0.13070  -3.401 0.000671 ***
## PredNL          -0.82250    0.15380  -5.348  8.9e-08 ***
## PredL           -1.73483    0.12140 -14.291  < 2e-16 ***
## ResLo:PredNL     0.38950    0.21120   1.844 0.065146 .
## ResLo:PredL     -0.07768    0.16566  -0.469 0.639134
## ---
## Signif. codes: 0 '***' 0.001 '**' 0.01 '*' 0.05 '.' 0.1 ' ' 1
##
## (Dispersion parameter for binomial family taken to be 1)
##
##      Null deviance: 1212.65  on 77  degrees of freedom
## Residual deviance:  641.86  on 72  degrees of freedom
## AIC: 955.54
##
## Number of Fisher Scoring iterations: 4
```

Wow, that is really overdispersed! If this was the approach you wanted to take, you can model an overdispersed binomial dataset with a type of model called a *beta-binomial*, using the function **betabin()** in the *aod* package ("analysis of overdispersed data"). However, the beta-binomial model is technically a type of mixed-effects or random-effects model, so we will discuss this technique in Chapter 8.

7.6 MIXING GLMS AND ANCOVAS TOGETHER

Let's make our model more interesting, at least for the point of illustrating the techniques we want to explore. Perhaps we want to know how survival to metamorphosis relates to the length of the larval period and predator treatment?

```
glm.negb2<-glm.nb(N.dead~Age.DPO*Pred, RxP.byTank)
summary(glm.negb2)
```

```
##
## Call:
## glm.nb(formula = N.dead ~ Age.DPO * Pred, data = RxP.byTank,
##     init.theta = 5.128645402, link = log)
##
## Deviance Residuals:
##     Min       1Q    Median       3Q       Max
## -2.6883   -0.8392   -0.1132   0.5432   2.7670
##
## Coefficients:
##                    Estimate Std. Error z value Pr(>|z|)
## (Intercept)        2.043076   0.308071   6.632 3.32e-11 ***
## Age.DPO            0.002329   0.003834   0.607 0.543607
## PredNL             1.559002   0.506024   3.081 0.002064 **
## PredL              1.953070   0.535289   3.649 0.000264 ***
## Age.DPO:PredNL    -0.016738   0.006883  -2.432 0.015024 *
## Age.DPO:PredL     -0.015103   0.009171  -1.647 0.099593 .
## ---
## Signif. codes: 0 '***' 0.001 '**' 0.01 '*' 0.05 '.' 0.1 ' ' 1
##
```

```
## (Dispersion parameter for Negative Binomial(5.1286) family
##   taken to be 1)
##
##      Null deviance: 168.971  on 77  degrees of freedom
## Residual deviance:  80.845  on 72  degrees of freedom
## AIC: 549.55
##
## Number of Fisher Scoring iterations: 1
##
##
##              Theta:   5.13
##          Std. Err.:   1.11
##
##   2 x log-likelihood:   -535.552
```

We can continue to use the **Anova()** function from the *car* package to calculate our summary statistics for the model.

```
Anova(glm.negb2)
```

```
## Analysis of Deviance Table (Type II tests)
##
## Response: N.dead
##              LR Chisq Df  Pr(>Chisq)
## Age.DPO         2.482  1     0.11519
## Pred           45.980  2  1.036e-10 ***
## Age.DPO:Pred    8.306  2     0.01571 *
## ---
## Signif. codes: 0 '***' 0.001 '**' 0.01 '*' 0.05 '.' 0.1 ' ' 1
```

Interestingly, this tells us that the number of tadpoles that died before metamorphosis was affected by "*Pred*," which we already knew, but also that the effect of predators was affected by how long tadpoles were in the water. Cool!

⊣ Box 7.6 - Model selection ⊢

As we've seen already, one way we can decide which model is the "best" is to use a function like **fitdistr()** to estimate how a response variable compares to different error distributions. But how do you determine which predictor variables to include? How do you determine which predictors, alone or in combination, best explain your response variable? Sure, the **Anova()** function can tell you which predictors are "significant" or not. But a technique called *model selection* provides an objective way to measure how well a set of predictors models a set a data. What you do is fit multiple models that have the same response variable but different predictors and evaluate their model fit with the function **AIC()**, just like we did when using the function **fitdistr()**. Any model in fit in R is given an AIC score, which is an objective measure of fit. Importantly, the AIC score takes in to account the number of predictors in your model, since having more predictors should theoretically allow you to explain the data better. AIC scores therefore put a value on explaining your data in the simplest yet most effective manner possible.

Before going further, here are the important things to always keep in mind with regards to model selection:

- You can only compare models fit to the *exact same* response variable.
- You can only compare models fit with the *exact same* error distribution.
- Models that have AIC scores within two of one another are generally considered not to be different.
- Lower scores are better.

In our previous example, maybe we want to know which combination of predictors best explains N.dead. Let's make a series of models and compare them with **AIC()**. Note that for the sake of brevity, I have not made every possible combination of the predictors (but you certainly can if you want).

```
glm.negb3<-glm.nb(N.dead~Res*Pred*Hatch, data=RxP.byTank)
glm.negb4<-glm.nb(N.dead~Res+Pred+Hatch, data=RxP.byTank)
glm.negb5<-glm.nb(N.dead~Res+Pred+Hatch+Pred:Res,
                  data=RxP.byTank)
glm.negb6<-glm.nb(N.dead~Res+Pred+Hatch+Pred:Res+Pred:Hatch,
                  data=RxP.byTank)
AIC(glm.negb3,glm.negb4,glm.negb5,glm.negb6)
```

```
##              df      AIC
## glm.negb3 13 554.2509
## glm.negb4  6 546.3987
## glm.negb5  8 549.8008
## glm.negb6 10 551.9161
```

What this shows us is that model "*glm.negb4*," the model with each individual predictor but no interactions at all, has the lowest AIC score, which gives us an indication that it does the best job of efficiently explaining the data. Thus, you could reasonably proceed with just including the three individual predictors in the model, ignoring the possible interactions.

⊣ Box 7.7 - Model reduction ⊢

A related concept to model selection is known as *model reduction*. Oftentimes we might code a model that has non-significant interactions and we just want to know the effect sizes of the significant predictors. In these cases, it can be useful to reduce your model to what is called the *minimal adequate model*. For example, in our model "*glm.negb*" earlier, we saw that the interaction between resources and predators was not significant. Thus, to get the most accurate measure of the effects of resources and predators, we could remove the interaction and re-run the model. Recall that the model we coded included the predictors "*Res*Pred*" which is a shortcut for "*Res+Pred+Res:Pred*." Thus, we want to remove the "*Res:Pred*" effect. There are two ways to do this. We can just code the individual effects of "*Res*" and "*Pred*" with a "+." We could also remove the interaction by writing "*Res*Pred-Res:Pred*." The second version is more useful if you have 3 or more predictors and just want to remove particular interactions.

```
glm.negb3<-glm.nb(N.dead~Res+Pred, data=RxP.byTank)
summary(glm.negb3)
```

```
##
## Call:
## glm.nb(formula = N.dead ~ Res + Pred, data = RxP.byTank,
##      init.theta = 4.555763667, link = log)
##
## Deviance Residuals:
##     Min       1Q    Median       3Q       Max
## -2.5562   -0.7835   -0.3639    0.4367    2.6892
##
## Coefficients:
##              Estimate Std. Error z value Pr(>|z|)
## (Intercept)    2.0884     0.1196  17.468  < 2e-16 ***
## ResLo          0.2414     0.1215   1.987  0.04693 *
## PredNL         0.4817     0.1756   2.743  0.00609 **
## PredL          1.1277     0.1352   8.341  < 2e-16 ***
## ---
## Signif. codes:  0 '***' 0.001 '**' 0.01 '*' 0.05 '.' 0.1 ' ' 1
##
## (Dispersion parameter for Negative Binomial(4.5558) family
```

```
##   taken to be 1)
##
##      Null deviance: 154.668  on 77  degrees of freedom
## Residual deviance:  80.163  on 74  degrees of freedom
## AIC: 551.72
##
## Number of Fisher Scoring iterations: 1
##
##
##                  Theta:   4.556
##             Std. Err.:   0.946
##
##   2 x log-likelihood:   -541.723
```

```
Anova(glm.negb3)
```

```
## Analysis of Deviance Table (Type II tests)
##
## Response: N.dead
##        LR Chisq Df Pr(>Chisq)
## Res      3.940   1    0.04716 *
## Pred    70.863   2  4.095e-16 ***
## ---
## Signif. codes:  0 '***' 0.001 '**' 0.01 '*' 0.05 '.' 0.1 ' ' 1
```

Since both "*Res*" and "*Pred*" are significant, we would dub this the minimal adequate model. Looking at the previous output, we can see that in this case it *does not* make a huge difference to remove the interaction. But, know that in some cases it can. Removing these higher-order interactions can sometimes change the estimation of significance of lower effects, so it is good to be very thorough. Generally, you want to start with the most complex version of your model you are interested (all the predictors and their interactions) and then prune the highest order interactions and re-run the model, continuing to prune interactions and predictors until you read the minimal adequate model.

7.7 USING THE **predict()** FUNCTION WITH A GLM

The model we made previously, "*glm.negb2*," is analogous to an analysis of variance (ANCOVA), as we have a mix of categorical and continuous predictors. We could try to figure out the nonlinear regression from the summary output, but instead we can use the very handy **predict()** function to calculate predicted values from the model. Remember that to use the **predict()** function, you need to provide a model and an empty data frame that has starter values for each of the predictors in your model. We will provide a sequence from 40–140 (the length of our x-axis), and our three Predator treatments. When you start to have multiple predictors that need to be accounted for, it can be very useful to use the **expand.grid()** function to build your blank data frame, which we have discussed before. **expand.grid()** will create a data frame for every combination of values you provide it. This can be very useful, but be careful, because you can easily create an enormous data frame!

Let's make an object called "*predicted.data*" where the output from **predict()** will be stored. Note that using **expand.grid()** in this manner does not work if you are trying to get confidence intervals (because that output is a list, which does not jive with a data frame). By looking at the head and tail of the object we can see what it has created. I will also show you a very easy way to plot confidence intervals shortly. I've also made a blank column here, where we will put our predicted values.

```
predicted.data<-expand.grid(Age.DPO=40:140,
                            Pred=c("C","NL","L"),
                            N.dead=NA)
head(predicted.data)
```

```
##    Age.DPO Pred N.dead
## 1      40   C      NA
## 2      41   C      NA
## 3      42   C      NA
## 4      43   C      NA
## 5      44   C      NA
## 6      45   C      NA
```

```
tail(predicted.data)
```

```
##      Age.DPO Pred N.dead
## 298     135   L      NA
```

```
## 299        136      L       NA
## 300        137      L       NA
## 301        138      L       NA
## 302        139      L       NA
## 303        140      L       NA
```

Now, we can use this new data frame to easily get our predicted values. Note that since we are using **predict()** with a glm, we want to include an argument to specify that the output is put back on the scale of our response variable, as opposed to the transformed scale of the model. In this case, that means the output will be in terms of tadpoles that died before metamorphosis.

```
predicted.data$N.dead<-predict(glm.negb2,
                               newdata=data.frame(predicted.data),
                               type="response")
head(predicted.data)
```

```
##     Age.DPO Pred   N.dead
## 1        40    C 8.467445
## 2        41    C 8.487187
## 3        42    C 8.506975
## 4        43    C 8.526810
## 5        44    C 8.546690
## 6        45    C 8.566617
```

If we wanted, we could plot these nonlinear curves from the negative binomial or do some calculations to estimate how different the effects of predators are at different points during the larval period. However, *ggplot2* makes it much easier to make that figure, so let's go ahead and do that.

7.8 MAKING A MUCH EASIER GLM/ANCOVA PLOT USING *ggplot2*

Now that you are a pro at using the **predict()** function, let me teach you a much easier way to make nice looking scatterplots with confidence intervals. We have previously seen that we can use **qplot()** to make quick and easy scatterplots. In model ***glm.negb2***, we are looking at the interaction between predators and age at metamorphosis. In Chapter 6, we saw how we can easily add curves to a plot in ***ggplot2*** by adding on the **geom_smooth()** function after our plot. You might recall that when we were making a linear regression, we specified "*method='lm'*". You might be able to guess how we'll modify that to plot the new curves. We'll just change the "*method=*" argument to *glm.nb*! So easy (see Figure 7.7).

```
qplot(x=Age.DPO, y=N.dead, data=RxP.byTank, col=Pred,
      ylab="Number of dead tadpoles",
      xlab="Age at metamorphosis (DPO)") +
  geom_smooth(method=glm.nb)+
  theme_classic()
```

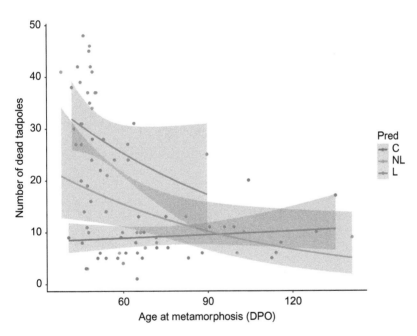

Figure 7.7 A scatterplot made with **qplot()**, which added nonlinear regression curves and their confidence intervals from a negative binomial regression. Wasn't that a piece of cake?

Box 7.8 - Take-home message

- GLMs are an extension of Linear Models (LMs) and are used to model non-normal data that fit one of several common error distributions. The coding is almost identical in both types of model, but GLMs include an argument to specify the error distribution family you are using.
- It is important to assess the diagnostic plots of your model residuals and the dispersion of the model in order to make an informed decision about which error distribution fits your data best and if you are meetig the assumptions of the model.
- Model selection and model reduction are techniques which allow you to find the most appropriate combination of predictors and/or the minimal adequate model.

7.9 ASSIGNMENT!

Here is an assignment to work on the skills you have just developed. Check the Github page (https://github.com/jtouchon/Applied-Statistics-with-R) for the book for sample solutions and code.

1. Explore how all three predictors from the experiment (*Hatch*, *Pred*, and *Res*) affect (or don't affect) the number of animals that died before metamorphosis. Start with the full model including all interactions and reduce it to the minimal adequate model, aka the model with just significant predictors.
2. Plot the data in an appropriate manner to visualize what significant affects you find.

Mixed Effects Models

8 Mixed Effects Models

The purpose of this chapter is to introduce you to how to code, diagnose, and interpret mixed effects models. It is assumed that you are using the **RxP**, **RxP.clean**, and **RxP.byTank** datasets created in earlier chapters.

8.1 UNDERSTANDING MIXED EFFECTS MODELS

Mixed effects models, also called random effects models, are a way to control for variation in your data that you are aware of but are not interested in. In essence, it allows you to partition variance in the data into two groups: the **Fixed effects** (what you are interested in studying) and the **Random effects** (sources of variation that may be important, but which are not things you are really interested in). One useful way to think about the difference between fixed and random effects is that fixed effects affect the *mean* of your data, whereas random effects primarily affect the *variance* of your data. Thus, by accounting for the random effects, you can reduce the noise in your data and get a clearer picture of the effects you are interested in.

Applied Statistics with R: A Practical Guide for the Life Sciences. Justin C. Touchon, Oxford University Press (2021). © Justin C. Touchon. DOI: 10.1093/oso/9780198869979.003.0008

Box 8.1 - What is a mixed effects model?

Mixed effects models are a form of what is called a *hierarchichal model*. Essentially, instead of just calculating a single regression (for example), the model builds regressions for each level of your random effects. Think of them as "sub-models." All of those sub-models are then pooled together to build a general effect (i.e., the fixed effect). The variation that exists between the sub-models is accounted for (i.e., the random effect), while preserving and potentially clarifying the fixed effect.

It is important to recognize that this is a mathematically complex process and if you have very complex random effects' structures or highly unbalanced random effects, the model may not run. For example, if you have very little data in one of the random effects, there might not be enough to build a model from, or the ability of the model to estimate the variance will be really poor. In any event, know that while running mixed effects models is in principle pretty straightforward, there are lots of things that can cause you headaches and hiccups.

One of the most important ways that mixed models can be used is that they allow you to use data from non-independent observations. For example, until now we have used data that were averaged for each mesocosm (e.g., mean time to metamorphosis, etc.). This is because each individual in a given tank is not independent from all the other individuals in the same tank. They had the exact same developmental environment! However, with mixed models we can use all the data from every individual while still controlling for the fact that metamorphs were raised in common environments.

Box 8.2 - There are many functions for running mixed models

There are multiple packages and functions for conducting mixed models. The most widely used and supported function is **lmer()** from the *lme4* package. The coding for **lmer()** is the same as for **lm()** and **glm()**, but with an added component to account for the random effects. You can also use most of the generalized linear model (GLM) error families with mixed models, which is very useful! To conduct generalized linear mixed models, use the function **glmer()** (also found in the *lme4* package). As with GLM's, there is a specific function for conducting a generalized linear mixed model with a negative binomial distribution, which is **glmer.nb()**. Other important packages that are useful to know about include **glmmadmb**, **glmmTMB**, and **MCMCglmm**. The first two (**glmmadmb** and **glmmTMB**) utilize different algorithms to estimate the model than **lme4**, and so you may not get the exact same answer under each function (but, hopefully they are very similar). **glmmadmb** in particular is much slower than **lme4** but is useful under conditions when **lme4** may fail to run, as is **glmmTMB**. Due to the different algorithms used, they are often more robust to things that cause problems in **lme4**. In recent years, **glmmTMB** has emerged as the better of the two. **glmmTMB** also allows you to run a set of potentially useful models called "zero-inflation" models. Having lots of zero's in your dataset can cause problems, particularly for count data distributions like Poisson and negative binomial, but zero-inflation models are built to handle them. The package **MCMCglmm** uses a Bayesian framework to estimate your parameters. Bayesian techniques are beyond the scope of this book and we will not discuss them here, but there are great resources and books on Bayesian methods available. In general, the functions within each of these packages to execute a model have the same name as the package (e.g., in **glmmTMB** you run a model with **glmmTMB()**).

8.1.1 Coding a mixed effects model

Using mixed models allows us to do things like explore the effect of resources and predators on metamorph snout-vent length (SVL), but use all of the data and control for the fact that metamorphs were raised in common tanks, and that tanks came from spatially separated blocks.

In order to add in the random effects, we simply tack on an additional argument in our model. After your fixed effects, you add additional effects inside sets of parentheses to specify what you want to include as random effects.

Box 8.3 - Types of random effects

R lets you construct two different types of random effects. Models with a *random intercept* are the most common and assume that the fundamental relationship between your predictors and response variables (i.e., the slope) is the same across the levels of your random effects. A *random slope* assumes instead that the fundamental relationship between predictor and response might be different across the levels of your random effects (see Figure 8.1).

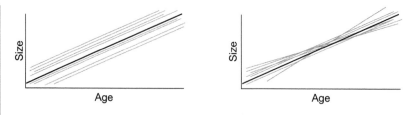

Figure 8.1 Coding random effects as a random intercept (left) means each of your random effects can have a different intercept, but the slopes will all be constrained to be the same. Coding both a random slope and intercept (right) will allow each random effect to vary more and represents when you think the random effect might have fundamentally changed the relationship between your predictor and response variables. In the two examples, the light gray lines represent the random effects, which pool together to create the fixed effect (the thick black line).

In order to code the random effect, you include an argument enclosed in parentheses, with the random slope on the left side of a pipe command "|" and the random intercept on the right side. Note that you cannot have a random slope model without a random intercept. Also, note that you can have multiple, *nested random intercepts* (this would normally be based on nested variables, where some effects fit entirely within another effect). It would look like this, where random_intercept2 is the effect operating *within* random_intercept1:

...+(random_slope|random_intercept1/random_intercept2)...

It is possible to have multiple, non-nested effects, also known as *crossed random effects*. This situation implies that multiple things are affecting the variance of your data, but those effects operate independent of one another. Crossed random effects look something like this:

...+(1|random_intercept_1)+(1|random_intercept_2)...

In the following model, we have included two nested, random intercepts, one for the spatial block that tanks were in and one for each individual tank that metamorphs came out of. The random effects are nested because metamorphs came out of tanks, and each tank was within a block. Thus, there are two spatial scales of effects, with tanks all being found within certain blocks. There is no random slope effect in these models because we are assuming that the fundamental relationship of how predators and resources combine to affect metamorph size is the same across all tanks. However, differences between blocks and tanks in proximity to the forest or overhead shade (just two possible sources of uncontrollable variation) might have had some effect on the size of froglets leaving the water. Lastly, please note that the column in the dataset "*Tank.Unique*" is just one where each tank is given a unique number instead of just being 1–12 in each block, which would cause some confusion (e.g., how can you have tank 1 in Block 1 and also in Block 2?). This model is the same as one we ran in Chapter 6, examining the effect of resources and predators, and their interaction, on the final age (days post-oviposition) of froglets. Here, however, we are using all the data instead of the tank means.

First, let's load all the packages we'll need.

```
library(dplyr)
library(ggplot2)
library(car)
library(lme4)
library(MASS)
library(emmeans)
```

```
#Code the lmm just like a regular lm,
#but with the random effect
lmm1<-lmer(log.Age.DPO~Res*Pred+(1|Block/Tank.Unique),
          data=RxP.clean)
summary(lmm1)
```

```
## Linear mixed model fit by REML ['lmerMod']
## Formula: log.Age.DPO ~ Res * Pred + (1 | Block/Tank.Unique)
##     Data: RxP.clean
##
## REML criterion at convergence: -152.3
##
## Scaled residuals:
##     Min      1Q  Median      3Q     Max
## -4.6111 -0.5561 -0.0161  0.6464  3.1204
##
```

```
## Random effects:
##  Groups              Name          Variance Std.Dev.
##  Tank.Unique:Block (Intercept) 0.038365 0.19587
##  Block             (Intercept) 0.008768 0.09364
##  Residual                      0.049453 0.22238
## Number of obs: 2493, groups: Tank.Unique:Block, 78; Block, 8
##
## Fixed effects:
##                  Estimate Std. Error t value
## (Intercept)      4.03758    0.05973  67.597
## ResLo            0.42107    0.07036   5.984
## PredNL          -0.03919    0.09311  -0.421
## PredL           -0.19079    0.07095  -2.689
## ResLo:PredNL    -0.09945    0.12809  -0.776
## ResLo:PredL     -0.26933    0.10096  -2.668
##
## Correlation of Fixed Effects:
##               (Intr) ResLo  PredNL PredL  RL:PNL
## ResLo         -0.588
## PredNL        -0.445  0.377
## PredL         -0.583  0.495  0.374
## ResLo:PrdNL    0.323 -0.549 -0.689 -0.272
## ResLo:PredL    0.410 -0.697 -0.264 -0.703  0.383
```

8.1.2 Interpreting the output

Notice that the output looks pretty much the same as what we saw for
lm() or **glm()**, however now we have a section titled "*Random effects*."
We have two random effects in this model, 1) Block and 2) Tank.Unique
nested within block. Looking at the "*Variance*" column tells us something
about the magnitude of our random effects, i.e., how much variation is
being accounted for purely by block and tank within block (written as
"*Tank.Unique:Block*" in the **summary()** output). However, there are no
good metrics for what is a "large" or "small" effect, so these are not
particularly easy to interpret. Remember though that part of the point is
that you don't really care about the variance of the random effects, you just
want to remove it and be done with it. The rest of the output is essentially the

same as we have seen before, and interpreting the coefficients is no different from **lm**() or **glm**().

8.1.3 Diagnostic plots

Because of the fact that *lme4* is a relatively new package, not all of the same utility functions exist that you can use with other, simpler models. For example, if you want to make the Q-Q plot, you have to either build it yourself or use a separate function that has been written in another package. To build it yourself, use the following code (Figure 8.2):

```
qqnorm(resid(lmm1));qqline(resid(lmm1))
```

Maybe now it is a good time to remind everyone what exactly a Q-Q plot is. In the previous code, we have first extracted the residuals of model *lmm1*, then plotted them with the **qqnorm**() function. If you think of an Analysis of Variance (ANOVA), the residuals are how different each of your data points are from the treatment mean, and if you think about a regression, it is how far the points are from the line of best fit. You can plot the residuals alone, along with a horizontal red line at 0, by typing:

```
plot(resid(lmm1))
abline(h=0, col="red", lwd=2)
```

What you see is that R has plotted how far from the predicted mean each point in our data frame is, from the 1st row the to 2493rd (Figure 8.3). If our model fits well, there will be an even spread of residuals above and below zero: most residuals should be close to zero, they should get fewer as you get farther away from zero, and they should be evenly split above and below zero. If you realize that a residual of zero *is the mean* then a normal distribution should have values evenly above and below the mean. Based on the size of our dataset and the basic spread of values, it would be feasible to calculate what a perfect set of residuals would look like. Thus, the Q-Q plot is just these hypothetical residuals (the x-axis, "*Theoretical Quantiles*") plotted against your real residuals (the y-axis, "*Sample Quantiles*"). If your data fit the model really well, then you would expect this to form a 1-to-1 line.

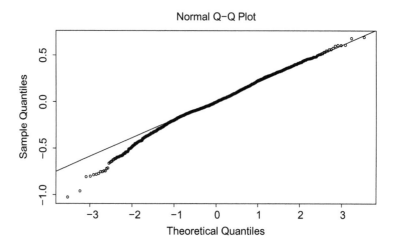

Figure 8.2 Q-Q plot for *lmm1*.

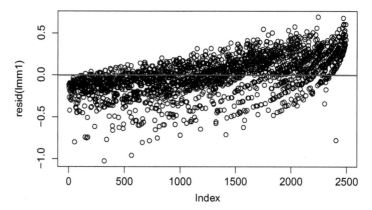

Figure 8.3 The residuals from model *lmm1*, with a red line indicating 0. Recall that a residual with a value of 0 means that value is exactly what the model would predict (i.e., it is the mean value).

In the past few years, several functions have arisen that will actually do this for you. In the **lattice** package, the **qqmath()** function will calculate the Q-Q plot (shown in Figure 8.4), and in the **sjPlot** package, the **sjp.lmer()** function will plot it as well, as long as you specify the argument "*type='re.qq.'*"

```
library(lattice)
qqmath(lmm1)
```

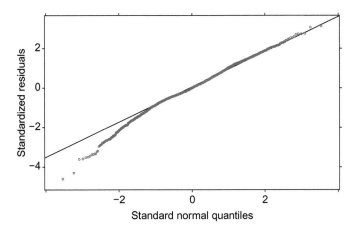

Standardized residuals

Standard normal quantiles

Figure 8.4 A different Q-Q plot for *lmm1*, made from the ***lattice*** package.

Box 8.4 - Obtaining p-values with mixed models: Likelihood ratio tests

With mixed effects models, you can use the **Anova()** function from the ***car*** package to calculate summary statistics, as we have done before, but it is not advised by the R gurus. Instead, you should use *Likelihood Ratio Tests* (LRT) to compare *nested models*. What does that mean? In essence, you build two models that differ in only one predictor and compare them. If the models are significantly different, that can be interpreted as statistical support for the predictor that differs between them. Our previous model, *lmm1*, contains three predictors: *"Res," "Pred,"* and the interaction between them—*"Res:Pred."* If we want to know, for example, if the interaction between *"Res"* and *"Pred"* is significant, we need to build a second model just like *lmm1* but without the interaction. We then use the function **anova()** to compare the two models. *Note that here we are using the version of the function with the lowercase "a." This is very important.* I know it is really confusing that there is not a separate function to conduct the LRT and that you use the same function **anova()**, even though you are not conducting an ANOVA. Such is life.

```
lmm2<-lmer(log.Age.DPO~Res+Pred+(1|Block/Tank.Unique),
          data=RxP.clean)
anova(lmm1,lmm2)
```

```
## refitting model(s) with ML (instead of REML)

## Data: RxP.clean
## Models:
## lmm2: log.Age.DPO ~ Res + Pred + (1 | Block/Tank.Unique)
## lmm1: log.Age.DPO ~ Res * Pred + (1 | Block/Tank.Unique)
```

```
##        npar     AIC      BIC logLik deviance  Chisq Df Pr(>Chisq)
## lmm2      7 -153.11 -112.36 83.554  -167.11
## lmm1      9 -156.44 -104.05 87.221  -174.44 7.3335  2    0.02556 *
## ---
## Signif. codes:  0 '***' 0.001 '**' 0.01 '*' 0.05 '.' 0.1 ' ' 1
```

The only difference between lmm1 and lmm2 is that lmm2 does not contain the interaction "*Res:Pred*." The **anova()** function compares the two models in a maximum likelihood framework and gives us a Chi-square statistic and a p-value, which can be viewed as the significance of "*Res:Pred*." Note that because of the hierarchical nature of mixed effects models, it is impossible to accurately know the degrees of freedom. Thus, you simply do not report them. This might seem strange if you have any experience with mixed effects models in another program, such as SAS, but this is the way R does it. SAS makes several assumptions in order to give you a concrete number for the degrees of freedom, but those assumptions are essentially unknowable and so R does not entertain such notions.

We can use LRT's to calculate p-values for all of our predictors. For example, we can compare our model lmm2 to models that contain only Pred or only Res and obtain p-values for all the predictors. Let's call them lmm3 and lmm4.

```
lmm3<-lmer(log.Age.DPO~Pred+(1|Block/Tank.Unique),
           data=RxP.clean)
lmm4<-lmer(log.Age.DPO~Res+(1|Block/Tank.Unique),
           data=RxP.clean)
anova(lmm2,lmm3) #P-value of Res
```

```
## refitting model(s) with ML (instead of REML)

## Data: RxP.clean
## Models:
## lmm3: log.Age.DPO ~ Pred + (1 | Block/Tank.Unique)
## lmm2: log.Age.DPO ~ Res + Pred + (1 | Block/Tank.Unique)
##        npar     AIC      BIC logLik deviance Chisq Df Pr(>Chisq)
## lmm3      6 -123.35  -88.42 67.674  -135.35
## lmm2      7 -153.11 -112.36 83.554  -167.11 31.76  1  1.745e-08 ***
## ---
## Signif. codes:  0 '***' 0.001 '**' 0.01 '*' 0.05 '.' 0.1 ' ' 1
```

```
anova(lmm2,lmm4)#P-value of Pred


## refitting model(s) with ML (instead of REML)

## Data: RxP.clean
## Models:
## lmm4: log.Age.DPO ~ Res + (1 | Block/Tank.Unique)
## lmm2: log.Age.DPO ~ Res + Pred + (1 | Block/Tank.Unique)
##       npar     AIC      BIC logLik deviance Chisq Df Pr(>Chisq)
## lmm4     5 -124.45  -95.341 67.224  -134.45
## lmm2     7 -153.11 -112.359 83.554  -167.11 32.66  2  8.088e-08 ***
## ---
## Signif. codes:  0 '***' 0.001 '**' 0.01 '*' 0.05 '.' 0.1 ' ' 1
```

Recall that in the version of this model that we ran in Chapter 6 (where we were using tank means), all three predictors ("*Res*," "*Pred*," and "*Res:Pred*") were significant, but the interaction between "*Res*" and "*Pred*" was very marginal at $P = 0.049$. Here, we see the same basic pattern of all three predictors being significant, but now we see an even stronger interaction between "*Res*" and "*Pred*." In other words, by accounting for and removing some of the variance due to blocks and tanks, we have purified the effect of our treatments of interest. By using all of the data while controlling for variation due to common environments with our random effects, we have increased our statistical power and thus our ability to detect differences. Yay mixed effects models!

8.1.4 GLMM's (and troubleshooting when they don't run)

Now, let's explore how mixed effects models can be used in a generalized linear model framework (hint: it's just like a linear model). In Chapter 7, we used a negative binomial model to understand the effects of our three treatment variables on the number of animals that died before metamorphosis. However, even though we were using tank-level values, those tanks come from spatially segregated blocks, and as we've already seen, the effects of those blocks might be important. As a first step, you might just plot the number of animals that died by block to see what sorts of patterns there are. Note that since "*Block*" is a numerical variable (1–8), it is helpful to convert it to a factor when plotting so that the boxplot comes out correctly. We can

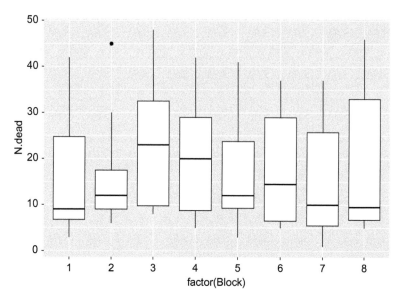

Figure 8.5 The number of tadpoles that died before metamorphosis in each block of the experiment.

do that easily by placing the variable inside the function **factor()** which changes it into a factor, but only will do so for this plot (see Figure 8.5).

```
qplot(data=RxP.byTank,
      x=factor(Block),
      y=N.dead,
      geom="boxplot")
```

That figure should make it clear that there is definitely some variation in survival across our blocks. There is substantial variation within each block, as noted by the distance between the top and bottom of each bar (which are the 25% and 75% percentage values), but also in the horizontal black lines, which are the medians in each block. The median mortality in Blocks 7 and 8 is about 10 tadpoles, whereas it is over 20 in Blocks 3 and 4.

Let's rerun our negative binomial model but as a mixed effects model including block as a random effect. The coding is identical to what you've done already, but now we are using the function **glmer.nb()**.

```
glmm1<-glmer.nb(N.dead~Pred*Res*Hatch+(1|Block),
                data=RxP.byTank)
```

Uh oh! Depending on what particular version of R and of *lme4* you are using, you might have received a bunch of warning messages when trying to run that model. Essentially, what all those warnings tell you is that the model is not being solved properly. Recall that a mixed effects model is a hierarchical model, which means that the function is trying to create a separate negative binomial regression for each block, then pool them together to get the overall fixed effects.

Maybe we should start more simply. Let's just run the model with the individual predictors, without any interactions, since that was the model that was significant in Chapter 6.

```
glmm2<-glmer.nb(N.dead~Pred+Res+Hatch+(1|Block),
                data=RxP.byTank)
```

Ugh, you might still be getting a warning that the model has failed to converge. What do you do now? We could ignore the warning. A lack of convergence just means the model didn't *converge* on a single best answer for what the effects parameters are. But it still came up with an answer and that might not be that far off from the "correct" answer. In fact, it might actually be the correct answer. Another option would be to try a different function, in this case **glmmTMB()**. A third option is to "go under the hood" a little bit and see if you can make the model run better. If you do a google search for the error, something like *"glmer Model failed to converge with max|grad"* you will find plenty of discussions where people have dealt with issues like this.

Suffice it to say, there are lots of things going on in the guts of the model that you can change. One thing that is particularly useful is to change the *optimizer* or the *number of iterations* the model runs. However, in this example that doesn't seem to help. Instead, we should try to figure out what the issue is. If we reduce the model down to just one predictor, will it run?

```
glmm3<-glmer.nb(N.dead~Pred+(1|Block), data=RxP.byTank)
summary(glmm3)
```

```
## Generalized linear mixed model fit by maximum likelihood
##    (Laplace Approximation) [glmerMod]
##  Family: Negative Binomial(4.5086)  ( log )
## Formula: N.dead ~ Pred + (1 | Block)
##    Data: RxP.byTank
##
##      AIC      BIC   logLik deviance df.resid
##    554.9    566.7   -272.5    544.9       73
##
## Scaled residuals:
##     Min      1Q  Median      3Q     Max
## -1.6719 -0.7332 -0.1993  0.5098  3.3974
##
## Random effects:
##  Groups Name         Variance Std.Dev.
##  Block  (Intercept) 0.01173  0.1083
## Number of obs: 78, groups:  Block, 8
##
## Fixed effects:
##              Estimate Std. Error z value Pr(>|z|)
## (Intercept)    2.2095     0.1094  20.190   <2e-16 ***
## PredNL         0.4482     0.1803   2.486   0.0129 *
## PredL          1.1303     0.1360   8.309   <2e-16 ***
## ---
## Signif. codes: 0 '***' 0.001 '**' 0.01 '*' 0.05 '.' 0.1 ' ' 1
##
## Correlation of Fixed Effects:
##         (Intr) PredNL
## PredNL -0.508
## PredL  -0.704  0.407
```

Okay! The only message we got was one saying that the fit was singular. In practice, that is probably not that big of a problem. It is just a message that a variance-covariance matrix in the estimation of the random effects is close to zero, i.e., it has no variance. This situation is not particularly uncommon and is likely not that big of a problem. Now what happens if we have two predictors?

```
glmm4<-glmer.nb(N.dead~Pred+Hatch+(1|Block), data=RxP.byTank)
summary(glmm4)
```

```
## Generalized linear mixed model fit by maximum likelihood
##    (Laplace Approximation) [glmerMod]
##  Family: Negative Binomial(4.9933)  ( log )
## Formula: N.dead ~ Pred + Hatch + (1 | Block)
##    Data: RxP.byTank
##
##      AIC      BIC   logLik deviance df.resid
##    550.6    564.7   -269.3    538.6       72
##
## Scaled residuals:
##     Min      1Q  Median      3Q     Max
## -1.8109 -0.6901 -0.1811  0.5125  3.3600
##
## Random effects:
##  Groups Name        Variance Std.Dev.
##  Block  (Intercept) 0.01115  0.1056
## Number of obs: 78, groups:  Block, 8
##
## Fixed effects:
##             Estimate Std. Error z value Pr(>|z|)
## (Intercept)   2.3486     0.1198  19.602   <2e-16 ***
## PredNL        0.4402     0.1739   2.532   0.0114 *
## PredL         1.1394     0.1313   8.678   <2e-16 ***
## HatchL       -0.3033     0.1182  -2.566   0.0103 *
## ---
## Signif. codes:  0 '***' 0.001 '**' 0.01 '*' 0.05 '.' 0.1 ' ' 1
##
## Correlation of Fixed Effects:
##        (Intr) PredNL PredL
## PredNL -0.460
```

```
## PredL  -0.606   0.412
## HatchL -0.465   0.011 -0.039
```

Once again, that worked okay. We still get the singularity message, but not any warnings. Do all combinations of two predictors allow the model to run?

```
glmm5<-glmer.nb(N.dead~Pred+Res+(1|Block), data=RxP.byTank)
glmm6<-glmer.nb(N.dead~Hatch+Res+(1|Block), data=RxP.byTank)
```

```
## boundary (singular) fit: see ?isSingular
```

So, we have an interesting conundrum here. We can't run the full model that we would want to run, but we can run parts of it. If the point of doing all this is to get an estimate of which predictors matter, we could run each model with a single pair of predictors and compare them against one another to estimate the significance of each individual effect. That seems less than ideal. Let's see what happens when we use a different package, namely *glmmTMB*. This is a pretty new and still somewhat experimental package, but it has been much more fully developed in recent years. Note that the package and function within the package have the same name. Also, notice that the coding is slightly different. Now, we have to add a "*family =*" argument, and in this case the family for a classic negative binomial error distribution is called "*nbinom2.*" Lastly, **glmmTMB()** requires our random effects to be factors (**lmer()** can convert this internally for us), so we have to code that ourselves.

```
#here is another package for running GLMMs
library(glmmTMB)
#First, convert Block to a factor
RxP.byTank$Block.factor<-as.factor(RxP.byTank$Block)
#Next, run the model. Note that the function and family
#specifications are different.
glmm7<-glmmTMB(N.dead~Pred+Res+Hatch+(1|Block.factor),
              family="nbinom2", data=RxP.byTank)
```

The model has successfully run with no errors! Let's run our LRT's to get our p-values.

```
#Create a series of models that only each differ
#in one predictor from one another
glmm7.1<-glmmTMB(N.dead~Pred+Res+(1|Block.factor),
                 family="nbinom2", data=RxP.byTank)
glmm7.2<-glmmTMB(N.dead~Pred+Hatch+(1|Block.factor),
                 family="nbinom2", data=RxP.byTank)
glmm7.3<-glmmTMB(N.dead~Res+Hatch+(1|Block.factor),
                 family="nbinom2", data=RxP.byTank)
#Run the likelihood ratio tests
anova(glmm7,glmm7.1)#sig of Hatch
```

```
## Data: RxP.byTank
## Models:
## glmm7.1: N.dead~Pred+Res+(1|Block.factor), zi=~0, disp=~1
## glmm7: N.dead~Pred+Res+Hatch+(1|Block.factor), zi=~0, disp=~1
##         Df   AIC    BIC logLik deviance Chisq Chi Df Pr(>Chisq)
## glmm7.1  6 552.96 567.10 -270.48   540.96
## glmm7    7 547.61 564.11 -266.80   533.61 7.3521      1   0.006699 **
## ---
## Signif. codes:  0 '***' 0.001 '**' 0.01 '*' 0.05 '.' 0.1 ' ' 1
```

```
anova(glmm7,glmm7.2)#sig of Res
```

```
## Data: RxP.byTank
## Models:
## glmm7.2: N.dead~Pred+Hatch+(1|Block.factor), zi=~0, disp=~1
## glmm7: N.dead~Pred+Res+Hatch+(1|Block.factor), zi=~0, disp=~1
##         Df   AIC    BIC logLik deviance Chisq Chi Df Pr(>Chisq)
## glmm7.2  6 550.55 564.69 -269.28   538.55
## glmm7    7 547.61 564.11 -266.80   533.61 4.9447      1   0.02617 *
## ---
## Signif. codes:  0 '***' 0.001 '**' 0.01 '*' 0.05 '.' 0.1 ' ' 1
```

```
anova(glmm7,glmm7.3)#Sig of Pred
```

```
## Data: RxP.byTank
## Models:
## glmm7.3: N.dead~Res+Hatch+(1|Block.factor), zi=~0, disp=~1
## glmm7: N.dead~Pred+Res+Hatch+(1|Block.factor), zi=~0, disp=~1
##         Df   AIC    BIC logLik deviance Chisq Chi Df Pr(>Chisq)
## glmm7.3  5 599.40 611.18 -294.7    589.40
## glmm7    7 547.61 564.11 -266.8    533.61 55.791      2  7.674e-13 ***
## ---
## Signif. codes:  0 '***' 0.001 '**' 0.01 '*' 0.05 '.' 0.1 ' ' 1
```

We get essentially the same answer as before, extremely significant effects of Pred, highly significant effects of Hatch, and significant effects of Res. The p-values are even very similar to what we got in Chapter 7, which tells us that in this case the block effects are minimal. Survival might have varied across blocks, but it wasn't in any appreciable way across the treatments and so does not distort or alter the overall effects we are seeing.

I wonder if the full model would run in **glmmTMB()**? Let's try it and see.

```
#Use glmmTMB to run the model
glmm8<-glmmTMB(N.dead~Pred*Res*Hatch+(1|Block.factor),
               family="nbinom2", data=RxP.byTank)
summary(glmm8)
```

```
##  Family: nbinom2   ( log )
## Formula:    N.dead ~ Pred * Res * Hatch + (1 | Block.factor)
## Data: RxP.byTank
##
##      AIC      BIC   logLik deviance df.resid
##    554.9    587.9   -263.4    526.9       64
##
## Random effects:
##
## Conditional model:
##   Groups       Name          Variance Std.Dev.
##   Block.factor (Intercept) 0.0147    0.1213
## Number of obs: 78, groups:  Block.factor, 8
##
## Overdispersion parameter for nbinom2 family (): 6.11
##
## Conditional model:
##                       Estimate Std. Error z value Pr(>|z|)
## (Intercept)            2.22020    0.18982  11.696  < 2e-16 ***
## PredNL                 0.77949    0.29853   2.611  0.00903 **
## PredL                  1.07479    0.24390   4.407 1.05e-05 ***
## ResLo                  0.26930    0.25455   1.058  0.29008
## HatchL                -0.48043    0.27654  -1.737  0.08233 .
## PredNL:ResLo          -0.54213    0.44423  -1.220  0.22232
## PredL:ResLo           -0.01835    0.33798  -0.054  0.95671
## PredNL:HatchL         -0.71831    0.49143  -1.462  0.14384
```

```
## PredL:HatchL           0.33411      0.35807    0.933  0.35077
## ResLo:HatchL           0.23974.     0.37591    0.638  0.52363
## PredNL:ResLo:HatchL    0.86164      0.67246    1.281  0.20008
## PredL:ResLo:HatchL    -0.30299      0.49211   -0.616  0.53809
## ---
## Signif. codes:  0 '***' 0.001 '**' 0.01 '*' 0.05 '.' 0.1 ' ' 1
```

Holy cow, it worked! The point here is that due to the different ways that different functions work, sometimes a model that might not run in one package will work just fine in another. It is always a good idea to start with **lmer()** or **glmer()** but if your data have issues, don't fret and try another function. It may not always help, but many times it will.

┤ **Box 8.5 - Notes of caution** ├

So, that's all there is to it! Mixed effects models are pretty easy, and they really are very useful. Here are a few things to keep in mind when using mixed models.

1. Don't include *too many random effects*. In any experiment it is probably possible to come up with a dozen different things that might have affected your data, but you can't try to include all of those in your model. The math that underlies a mixed model is complicated, and if you try to create too complex of a random effects structure your model won't run.
2. Try to keep your random effects balanced. What that means is that if you specify random effects that only apply to one treatment (for example) or one particular subset of the data, then your model might not run. This really emphasizes the importance of experimental design. You want your experimental units to be evenly distributed across the levels of your random effects, to whatever extent is possible.
3. You may not need a mixed model in the first place. If you don't need it, don't bother with it. Sometimes you just don't need to include random effects at all.

8.1.5 More about model selection

How do you know if you should include some random effects in your model? One way is by using Akaike's Information Criterion (AIC) to compare two models that differ only in the fact that one has random effects and the other does not. For example, we can build a linear model that is just like *lmm1* but doesn't have the random effects. In fact, we already did that in Chapter 6. It was called *lm2*. Here, however, we will use *RxP.clean* in order to facilitate comparison with *lmm1*.

```
lm6<-lm(log.Age.DPO~Res*Pred, data=RxP.clean)
```

Now, one important thing to know is that ***lme4*** uses a technique to fit mixed effects models called *restricted maximum likelihood*, or (REML) for short. Regular linear models and generalized linear models are fit with a standard maximum likelihood approach. I won't get into the difference between them here but know that if you want to compare a linear model and linear mixed effects model, you have to fit the mixed model *without using REML*. That is easy enough, you just specify "*REML = F*" in the model. As a side note, you might have noticed a message earlier when we used **anova**() to calculate likelihood ratio tests that stated, "*refitting model(s) with maximum likelihood (ML) (instead of REML).*" What this was saying is that to compare models we shouldn't be using REML, which is how the models were originally fit, so R took the liberty of refitting our models using maximum likelihood in order to facilitate the likelihood ratio test. Thanks R!

```
#Create a mixed model, but without using REML
lmm1<-lmer(log.Age.DPO~Res*Pred+(1|Block/Tank.Unique),
           data=RxP.clean, REML = F)
```

Now, we can compare these two models using the function **AIC**(). We have previously used AIC to compare the fit of a dataset to different error structures and the idea of model selection was introduced in Chapter 7. Much like model selection, here we are going to use it to compare the fit of two models against one another. Recall that AIC scores are unitless and that lower AIC scores mean a "better fit." In other words, a lower AIC score indicates the model does a better job of explaining the variance in the data.

```
#Compare two models, one with random effects
#and one without
AIC(lmm1,lm6)
```

```
##         df         AIC
## lmm1    9  -156.4413
## lm6     7  1113.8245
```

We can clearly see that model lmm1 which includes random effects for blocks and tanks within blocks has a much, much lower AIC score. Thus, we should keep the random effects in the model. We can see that the block has a large impact on the results if we plot our two predictors at the same time and facet the plot by block. Some blocks have a very strong interaction between resources and predators (e.g., Block 1) whereas other do not (e.g., Blocks 2 or 3). Moreover, the basic patterns are not entirely consistent across blocks. For example, in Block 1 the effect of lethal predators is much stronger in the high resource environment, but this is not the case on Block 2 or 8 (Figure 8.6).

```
qplot(y=log.SVL.final,
      x=Pred,
      fill=Res,
      data=RxP.clean,
      geom="boxplot",
      facets=.~Block)
```

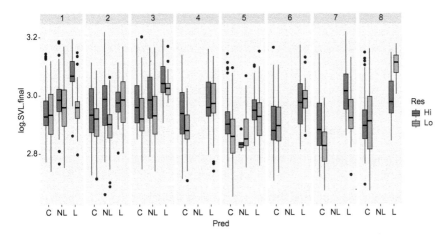

Figure 8.6 A series of boxplots showing how different the effects of predators and resources were in each block.

We can also use **AIC()** to compare models that differ in their random effects structures. For example, maybe we want to know if it is necessary to include the Tank.Unique effect nested inside the block effect. We can fit another model and compare them:

```
lmm1.1<-lmer(log.Age.DPO~Pred*Res+(1|Block),
             data=RxP.clean, REML = F)
AIC(lmm1,lmm1.1)
```

```
##          df      AIC
## lmm1      9 -156.4413
## lmm1.1    8  767.2936
```

Once again, we can see that including the second random effect dramatically lowers the AIC score, meaning we should definitely keep it in the model. We can even use **anova()** to estimate significance of the random effect here, although we generally are not interested in statistical significance of random effects. Given the huge difference in AIC scores, we might expect that it will be significant.

```
anova(lmm1,lmm1.1)
```

```
## Data: RxP.clean
## Models:
## lmm1.1: log.Age.DPO ~ Pred * Res + (1 | Block)
## lmm1: log.Age.DPO ~ Res * Pred + (1 | Block/Tank.Unique)
##        npar    AIC     BIC  logLik deviance  Chisq Df Pr(>Chisq)
## lmm1.1    8  767.29  813.86 -375.65   751.29
## lmm1      9 -156.44 -104.05   87.22  -174.44 925.73  1  < 2.2e-16 ***
## ---
## Signif. codes:  0 '***' 0.001 '**' 0.01 '*' 0.05 '.' 0.1 ' ' 1
```

Yup, we were right!

8.1.6 Estimating marginal means with mixed models
Due to the fact that mixed effects models account for, and remove, variation in your data due to the specified random effects, you can obtain model estimates that differ from your raw data. If your random effects had a particularly strong influence on your data, these differences can be substantial.

Why does this matter? It can be very important when plotting your data! As much as possible, you want to build graphics that are representative of the data analysis, and if controlling for random effects renders your treatments significant or demonstrably changes the estimates of treatment means, you want your figures to reflect this.

Luckily, this is easily accomplished by estimating marginal means via the **emmeans** package. First, let's see an example of this. Let's calculate the means of final metamorph SVL across our three predator treatments from our raw data, and then see what the predicted means are after controlling for "*Block*" and "*Tank*" effects. In the following code, I've used the function **fixef()** to access just the coefficients of the fixed effects instead of typing in the coefficients themselves. Since we know the order of different levels of "*Pred*," and we know how the parameters are built in the summary output, we can just exponentiate and add together the different fixed effects to get the predicted value of each parameter without having to type the numbers in ourselves.

```
#First, calculate the means from the raw data
RxP.clean %>%
  group_by(Pred) %>%
  summarize(mean.SVL.final = mean(SVL.final))
```

```
## # A tibble: 3 x 2
##    Pred  mean.SVL.final
##    <fct>          <dbl>
## 1 C               18.5
## 2 NL              19.2
## 3 L               19.8
```

```
#Next, run a mixed model analyzing log.SVL.final
lmm1<-lmer(log.SVL.final ~ Pred+(1|Block/Tank.Unique),
           data=RxP.clean)

#We can view just the parameters with the fixef() function
#Calculate Control treatment size
exp(fixef(lmm1)[1])
```

```
## (Intercept)
##     18.37956
```

```
#Calculate Nonlethal treatment size
exp(fixef(lmm1)[1]+fixef(lmm1)[2])
```

```
## (Intercept)
##     18.77424
```

```
#Calculate Lethal treatment size
exp(fixef(lmm1)[1]+fixef(lmm1)[3])
```

```
## (Intercept)
##     20.20379
```

Comparing the raw means with the estimated means from the mixed model is quite informative. After controlling for variation due to the blocks and tanks, our predicted mean SVL for metamorphs has decreased for tadpoles in the Nonlethal treatment by about 0.5 mm but increased for the Lethal treatment by about 0.4 mm. In other words, the model is telling us that the effect of lethal predators *is even larger* than we might think by looking at the raw data. If we want to make a figure that accurately plots these differences, we need to use **emmeans()** to predict the means, and their standard errors or confidence intervals, back out of the model. We previously used *emmeans* to do post-hoc tests, and we could do that again. But here, we can use it to obtain our predicted means with predicted standard error (the "*SE*" column) or a 95% confidence interval (the "*lower.CL*" and "*upper.CL*" columns). How wonderful! You could then easily exponentiate these values to get accurate values for a bar graph and error bars.

```
ph1<-emmeans(lmm1, specs="Pred")
summary(ph1)
```

```
##  Pred emmean     SE   df lower.CL upper.CL
```

```
## C      2.91 0.0124 11.2    2.88    2.94
## NL     2.93 0.0164 25.7    2.90    2.97
## L      3.01 0.0127 12.4    2.98    3.03
##
## Degrees-of-freedom method: kenward-roger
## Confidence level used: 0.95
```

Box 8.6 - Take-home message

- Mixed effects models allow you to partition the variance in your data into two groups: that which you are interested in, called the fixed effects, and that which you are generally not interested in, called the random effects.
- Fixed effects influence the mean of your data whereas random effects influence the variance.
- There are many functions to run mixed effects models in R. The most widely used are **lmer()** and **glmer()** in the **lme4** package.
- Coding a mixed effects model is essentially the same as a linear model or generalized linear model, with the exception that you include your random effects specified inside a set or sets of parentheses.
- Creating overly complicated random effects' structures may prevent your model from running. Try to keep your random effects as simple as necessary.
- Estimating marginal means with the **emmeans** package can be very useful with mixed effects models, as accounting for the variance due to random effects can sometimes greatly alter the estimates of your treatment means or regression coefficients.

8.2 ASSIGNMENT!

Here is an assignment to work on the skills you have just developed. Make sure to check the Github page (https://github.com/jtouchon/Applied-Statistics-with-R) for sample code and answers to these questions.

1. Rerun the ANCOVA models you explored in Chapter 6 but using all of the data and include random effects for block and tank within block. Don't forget to check the model fit via the Q-Q plot. How does it look?

2. Compare the AIC scores of these models with and without the random effects. Note that this is not a comparison of versions of the models from Chapter 6, which used tank means, and your new model. There is no way to justly compare those since they have different response variables (one being each individual tadpole and the other being the tank means of those individuals). Instead, you are making a model like that in the previous section titled "More about model selection," where you include all the data but without the random effects.

3. For each model you run, what effect does including all the data and adding in random effects have on our results? Why do you think this is?

Advanced Data Wrangling and Plotting

9 Advanced Data Wrangling and Plotting

The purpose of this chapter is to expand your skills with data manipulation and plotting techniques using packages such as *dplyr*, *ggplot2*, and more.

9.1 THE "TIDYVERSE"

Hadley Wickham is the author of a great number of very useful packages for "data wrangling," aka, highly functional and elegant data manipulation. He is also the chief data scientist for RStudio and has created a coding ecosystem referred to as the *tidyverse*. His books about data manipulation in the *tidyverse* ("**R for Data Science**") and about plotting with *ggplot2* ("**ggplot2: Elegant Graphics for Data Analysis**") are both excellent and can be read for free online. That said, I recommend purchasing hard copies, ideally through your local bookstore, but more likely on the internet.

For our purposes, there are three main packages within the *tidyverse* we are interested in. There are several others that will load too, but these three are the most important for what we are doing.

Applied Statistics with R: A Practical Guide for the Life Sciences. Justin C. Touchon, Oxford University Press (2021). © Justin C. Touchon. DOI: 10.1093/oso/9780198869979.003.0009

- **ggplot2**: Greatly enhanced data visualisation
- **dplyr**: Improved data manipulation
- **tidyr**: Data tidying

Very handily, you can load the whole collection of packages just by typing:

```
library(tidyverse)

## -- Attaching packages -------------------- tidyverse 1.3.0 --
## v ggplot2 3.3.2     v purrr   0.3.4
## v tibble  3.0.1     v dplyr   1.0.0
## v tidyr   1.1.0     v stringr 1.4.0
## v readr   1.3.1     v forcats 0.5.0
## -- Conflicts ----------------------- tidyverse_conflicts() --
## x dplyr::filter() masks stats::filter()
## x dplyr::lag()    masks stats::lag()
## x dplyr::select() masks MASS::select()
```

Depending on what other packages you have currently loaded, you might have gotten some warning messages like those shown above saying that some functions have been "masked." This happens when two different packages have functions with the same name. For example, the **MASS** package and the **dplyr** package each have a function called **select()**. We will need to use this function later, but how will R know which version to use?

When you load a package with a function that is already currently present in the workspace, the function that was loaded second will not be used by default. Thus, as the *"Conflicts"* section in the output says, the version of **select()** from **dplyr** is being *masked* by the version in **MASS**, which was already present. The solution is that you can specify which package you want a function to come from by typing the name of the package and two colons before the name of the function. For example, if you want to specify you are using the version of **select()** from **dplyr**, you can type the following:

```
dplyr::select()
```

Another very handy package is **broom**, which converts slightly messy objects like model outputs (which are often lists) in to "tidy" objects which can more easily be navigated.

```
library(broom)
```

Box 9.1 - Important functions to know in the *tidyverse*

There are many, many functions in any given package. Here are the most useful ones for data wrangling in the **tidyverse** (some of which you hopefully remember from earlier chapters).

- **%>%**: The pipe function takes the output of one line and makes it the input to the next line. This is extremely useful for stringing together sets of functions.
- **group_by()**: Allows you to create a grouped data object. It doesn't change anything about the how the data are stored, but it gives R the knowledge that you have grouping variables you are interested in, and those groups are inherited by subsequent commands.
- **select()**: Allows you to include or exclude certain variables (columns) from a dataset.
- **filter()**: Allows you to select just certain rows based on specified criteria. It is how you subset your data in the tidyverse.
- **summarize()**: How you calculate summary calculations, such as group means or standard deviation, for example.
- **mutate()**: How you calculate new variables (columns) from existing ones.
- **join**: How do you combine two different data objects into a single one? By "joining" them! There are many types of "joins" you can do, based on the specific type of data you want to keep or exclude (these will be explained in detail later on). Note that the two objects you are combining must share some common column which allows the data to be matched up.
- **do()**: This allows you to execute specific functions, such as running a model on every partition of a grouped dataset.
- **spread()**: This allows you to take a long dataset and spread it out wide.
- **gather()**: This does the opposite of spread, it takes a wide dataset and makes it long.

If some of these definitions do not make complete sense to you right now, hopefully they will once we start putting them in to practice. Also, note that there are many more useful functions than just these at your disposal in the tidyverse. These are just the most important and useful to know.

9.2 BASIC DATA WRANGLING

We will begin exploring how to summarize your data in various ways, including calculating new variables as necessary. We have done some of this type of thing already, so hopefully this serves as a refresher, reinforcement, and expansion of material already introduced in earlier chapters.

Calculating treatment means is one of the most common things a scientist might need to do. This is useful for plotting and also simply for knowing the average values across different categories of your data. What is the average effect of your experimental treatments? How tall are the plants in each species you have collected? What is the level of expression for each of the genes in your RNA-seq dataset? These are the sorts of things you can start to answer only once you summarize your data in some way.

Let's start by calculating the mean age at emergence (in terms of days post-oviposition) for each of our predation treatments. This code has 3 steps: 1) We begin by declaring our original dataframe (*RxP.clean*), 2) then we set our grouping variable, and 3) lastly, we define the new variable to calculate. Note that at each step of this process we **pipe** one line to the next using the **%>%** function.

9.2.1 Group and summarize data

```
#Here is an example of how you can summarize a variable
RxP.clean %>% #Define the starting data frame
  group_by(Pred) %>% #Define how to group the data
  summarize(Mean.Age.DPO = mean(Age.DPO)) #Calculate the mean
```

```
## # A tibble: 3 x 2
##    Pred  Mean.Age.DPO
##    <fct>        <dbl>
## 1 C             75.3
## 2 NL            73.3
## 3 L             53.5
```

Okay, let's take a moment to notice a few things. First, since you've designated the data frame at the outset, you can just refer to individual columns within that data frame without the need to use the $ operator. Secondly, here we didn't need to **select()** or **filter()** our data frame. The only things that are kept in our data frame are the grouping variable and the newly calculated mean values. Lastly, we didn't assign this to be a new object, so it just ran in the console and was not stored. We can easily add more grouping variables, more columns to summarize, or summarize the same column in different ways.

Just as an example, let's calculate the means of multiple columns.

```
#Here is an example of how you can summarize
#multiple variables in the same way
RxP.clean %>%
   group_by(Pred) %>%
   summarize(Mean.Age.DPO = mean(Age.DPO),
             Mean.Mass.final = mean(Mass.final),
             Mean.Resorb.days = mean(Resorb.days))
```

```
## # A tibble: 3 x 4
##    Pred  Mean.Age.DPO Mean.Mass.final Mean.Resorb.days
##    <fct>        <dbl>           <dbl>            <dbl>
## 1 C             75.3           0.366             4.32
## 2 NL            73.3           0.408             4.37
## 3 L             53.5           0.454             4.27
```

Okay, now let's calculate the means, standard deviations, and sample sizes of age at emergence for all combinations of the predator, resource, and hatching age treatments. These examples show how you can calculate different values from the same variable or calculate the same type of value from different variables. In this instance, let's go ahead and assign this output to be an object using the <-.

```
#Now, we are summarizing the
#same variable in different ways
RxP.summary <- RxP.clean %>%
   group_by(Pred, Res, Hatch) %>%
   summarize(Mean.Age.DPO = mean(Age.DPO),
             SD.Age.DPO = sd(Age.DPO),
             N.Age.DPO = length(Age.DPO))
RxP.summary
```

```
## # A tibble: 12 x 6
## # Groups:   Pred, Res [6]
##     Pred  Res   Hatch Mean.Age.DPO SD.Age.DPO N.Age.DPO
##     <fct> <chr> <chr>       <dbl>      <dbl>     <int>
##  1 C     Hi    E           55.3       14.2       325
##  2 C     Hi    L           61.8       16.2       354
##  3 C     Lo    E           99.2       39.8       302
##  4 C     Lo    L           87.9       28.5       324
##  5 NL    Hi    E           58.4       13.4       118
##  6 NL    Hi    L           56.8       12.5       131
##  7 NL    Lo    E           79.0       31.8       101
##  8 NL    Lo    L           96.4       42.4       144
##  9 L     Hi    E           48.9        9.49      184
## 10 L     Hi    L           47.5        6.20      214
## 11 L     Lo    E           59.5       23.8       122
## 12 L     Lo    L           61.4       18.3       174
```

9.2.2 Use **mutate()** to calculate new variables

With these data, we can easily use the **mutate()** function to calculate the standard error of the age at emergence, which is useful in plotting. Recall that standard error is calculated as the standard deviation divided by the square root of your sample size. In the following code, we pipe the output of our **summarize()** function to the **mutate()** function. By the time the data object gets to the mutate step, the columns "*SD.Age.DPO*" and "*N.Age.DPO*" exist, so we can refer to them in the calculation of SE. This last point is really, really important. As we move down the string of commands, we have to remember that we are not really working with the original starting data frame any longer. At each step we are modifying the data frame in some way, sometimes creating a whole new data frame which contains new variables with different names. We can use those variables to do new operations, even though they did not exist at the outset of the code.

```
RxP.summary <- RxP.clean %>%
  group_by(Pred, Res, Hatch) %>%
  summarize(Mean.Age.DPO = mean(Age.DPO),
            SD.Age.DPO = sd(Age.DPO),
```

```
              N.Age.DPO = length(Age.DPO)) %>%
  mutate(SE.Age.DPO = SD.Age.DPO/sqrt(N.Age.DPO)) #New variable!
RxP.summary
```

```
## # A tibble: 12 x 7
## # Groups:   Pred, Res [6]
##    Pred Res  Hatch Mean.Age.DPO SD.Age.DPO N.Age.DPO SE.Age.DPO
##    <fct> <chr> <chr>      <dbl>      <dbl>     <int>      <dbl>
##  1 C    Hi   E           55.3       14.2       325      0.790
##  2 C    Hi   L           61.8       16.2       354      0.860
##  3 C    Lo   E           99.2       39.8       302      2.29
##  4 C    Lo   L           87.9       28.5       324      1.58
##  5 NL   Hi   E           58.4       13.4       118      1.23
##  6 NL   Hi   L           56.8       12.5       131      1.09
##  7 NL   Lo   E           79.0       31.8       101      3.16
##  8 NL   Lo   L           96.4       42.4       144      3.54
##  9 L    Hi   E           48.9        9.49      184      0.699
## 10 L    Hi   L           47.5        6.20      214      0.423
## 11 L    Lo   E           59.5       23.8       122      2.15
## 12 L    Lo   L           61.4       18.3       174      1.39
```

9.2.3 Use **select()** to include or exclude certain variables

This is all so easy and cool right? What if you want to just calculate the means of everything in your dataset? That's easy. There is a specific function called **summarize_all()** which does just that. All you have to do is supply the function you want to apply. Now, it doesn't make sense to try to calculate the mean of *every* variable in the *RxP.clean* dataset, so let's use **select()** to remove the columns that we don't want to summarize. To remove columns from consideration, just place a minus sign in front of the column name. Also, since we might have a problem with **select()** getting masked by the **MASS** package, let's make sure to specify that we want to use **select()** from the **dplyr** package.

```
RxP.clean %>%
  dplyr::select(-Ind, -Block, -Tank, -Tank.Unique, -Hatch, -Res)
  %>%
  group_by(Pred) %>%
  summarize_all(mean) #summarize everything at once
```

```
## # A tibble: 3 x 11
##   Pred Age.DPO Age.FromEmergen… SVL.initial Tail.initial
##   <fct>   <dbl>            <dbl>       <dbl>        <dbl>
```

```
## 1 C          75.3              41.3        18.1        5.55
## 2 NL         73.3              39.3        18.9        5.68
## 3 L          53.5              19.5        19.4        5.84
## # … with 6 more variables: SVL.final <dbl>, Mass.final <dbl>,
## #    Resorb.days <dbl>, log.SVL.final <dbl>,
## #    log.Age.FromEmergence <dbl>, log.Age.DPO <dbl>
```

9.2.4 Use **filter()** to remove certain values

In the experiment that we are working with, there were originally eight experimental blocks of tanks. Imagine for a minute that we've been looking through our field notes and discovered that there might be an error in Block 4, such that now we want to know if the means differ substantially with and without that block. We can use **filter()** to remove just the rows from the dataset that correspond to Block 4. This is the same as using square brackets ([ ]) to subset a data frame. Regardless of which technique you use, you will need to set up a logical statement to define which values to keep or exclude. Logical statements were discussed in Chapter 3, so I won't go over them again here.

```
RxP.no4.summary <- RxP.clean %>%
  filter(Block != 4) %>%
  group_by(Pred, Res, Hatch) %>%
  summarize(Mean.Age.DPO.no4 = mean(Age.DPO))
RxP.no4.summary
```

```
## # A tibble: 12 x 4
## # Groups:   Pred, Res [6]
##      Pred  Res    Hatch Mean.Age.DPO.no4
##      <fct> <chr> <chr>            <dbl>
## 1 C        Hi     E                53.7
## 2 C        Hi     L                58.7
## 3 C        Lo     E                96.9
## 4 C        Lo     L                86.2
## 5 NL       Hi     E                58.4
## 6 NL       Hi     L                56.8
## 7 NL       Lo     E                79.0
## 8 NL       Lo     L                96.4
```

##	9 L	Hi	E	49.6
##	10 L	Hi	L	47.5
##	11 L	Lo	E	51.7
##	12 L	Lo	L	62.1

9.2.5 Join two datasets together

Okay, now we have two different objects that we might like to compare. The first is *RxP.summary* and contains the calculations for age at emergence including all the blocks in the experiment, and the other is *RxP.no4.summary*, which contains the mean calculated without Block 4. Let's join those two objects to make it easier to compare the values. Before we do that, know that you have to have something in common in the two objects that will serve as a **key** to match the data objects together. Also know that there are *seven* different ways to join two data frames together (four of which we will discuss here), and yet another *two* ways to code it. The first object you name is your "*x*" data frame, and the second is the "*y*" data frame.

Box 9.2 - Types of joins

1. **full_join()**: This keeps all the columns from both data frames and fills in NA's anywhere there isn't a perfect match. This is useful, as it ensures that nothing is discarded from either dataset you are working with.
2. **left_join()**: This keeps all the columns from both data frames, but only the rows where there is a match to x from y. If there is a value in your y data frame that doesn't match up with your x, it will be dropped from the output.
3. **right_join()**: As you might guess, this is just the opposite of the left join. It keeps all the columns but only the rows where there is a match to y from x. Left and right joins are therefore identical, it just depends on which data frame you choose to call x versus y.
4. **inner_join()**: Once again, this keeps all the columns but only the rows where there is a match between both data frames. Any rows with missing values are dropped. This type of join will create no NAs, since every value kept has a match in both datasets.

It is useful to know that you can either code the join function by specifying your x and y data objects or you can pipe the x data object to the join function, which then specifies your y object. In either case, you need to specify the column or columns to match with a "*by=*" argument. Under most circumstances, you will likely just join based on a single column, the name of which has to be specified in quotes. Here, we have three columns which are necessary to properly align the values, so we have to concatenate them together into a single vector, which is quite useful to know how to do!

```
#These do the exact same thing.
#Declare your x and y data inside the join
full_join(x = RxP.no4.summary,
          y = RxP.summary,
          by=c("Pred","Res","Hatch"))
```

```
#Declare your x data, then pipe it to the join
RxP.no4.summary %>%
    full_join(RxP.summary,
              by=c("Pred","Res","Hatch"))
```

```
## # A tibble: 12 x 8
## # Groups:   Pred, Res [6]
##     Pred Res   Hatch Mean.Age.DPO.no4 Mean.Age.DPO SD.Age.DPO
##     <fct> <chr> <chr>            <dbl>        <dbl>      <dbl>
##  1 C     Hi    E                 53.7         55.3       14.2
##  2 C     Hi    L                 58.7         61.8       16.2
##  3 C     Lo    E                 96.9         99.2       39.8
##  4 C     Lo    L                 86.2         87.9       28.5
##  5 NL    Hi    E                 58.4         58.4       13.4
##  6 NL    Hi    L                 56.8         56.8       12.5
##  7 NL    Lo    E                 79.0         79.0       31.8
##  8 NL    Lo    L                 96.4         96.4       42.4
##  9 L     Hi    E                 49.6         48.9        9.49
## 10 L     Hi    L                 47.5         47.5        6.20
## 11 L     Lo    E                 51.7         59.5       23.8
## 12 L     Lo    L                 62.1         61.4       18.3
## # ... with 2 more variables: N.Age.DPO <int>, SE.Age.DPO <dbl>
```

Now, we can easily compare the two sets of means with and without Block 4. In general, it looks like excluding Block 4 has very little effect on the average age at emergence. The one exception is the Lethal-Low-Early treatment, which had a change of about 8 days, which is pretty substantial. If we wanted, maybe we should explore this rabbit hole a little more and see why that is the case. We won't do that here, but we could if we wanted to.

9.3 ADVANCED DATA WRANGLING: SPREADING AND GATHERING YOUR DATA

As we have seen here and in earlier chapters, we can calculate means or other summary statistics on our dataset by adding row after row to our **summarize()** function. But, when we are doing the same calculation on many columns, it can be more useful to use **gather()** and **spread()** to change the shape of our dataset. These two functions will come in very handy to you I am sure!

Box 9.3 - What is gathering or spreading?

To start off, let's discuss what "gathering" your data means. Essentially, we want to take the data that are in a bunch of columns and put them in to two columns, one that contains what used to be the column names and the other which contains the data that were in the columns. Thus, you end up with one column of categorical data, which is called the **key** (the former column headings), and one of numerical data, called the **value** (the values that were in each column). "Spreading" your data is just the opposite. You take two columns, one of categories/factors and one of numbers and place them into multiple columns. Importantly, in both of these steps you have the capability to select just the columns you want to spread or gather.

First, let's see what it looks like to gather our data into a super long format. First, we will pipe our dataset to the **gather()** function. Then, we will define the names of the key and value columns. For our example, let's call the key "*Measurement*" since the column headings represent different measurements that were taken on each metamorph and we'll call the value simply "*Value*," since each of the numerical values is the value recorded for a particular measurement. Lastly, we will select out all the columns *we don't want to gather*. We could just as easily do this with the **select()** function. It is just a question of personal preference. Note that this will create an *extremely long* data frame, so we can look at the head and tail of the object to get an idea of what it looks like.

```
RxP.clean.long <- RxP.clean %>%
   gather(key = Measurement, value = Value,
```

```
              -Ind, -Block, -Tank, -Tank.Unique, -Hatch, -Pred, -Res)
head(RxP.clean.long)
```

```
##  Ind Block Tank Tank.Unique Hatch Pred Res Measurement Value
## 1   1     5    7          55     E   NL  Hi     Age.DPO    35
## 2   2     5    4          52     L    C  Hi     Age.DPO    35
## 3   3     5    4          52     L    C  Hi     Age.DPO    35
## 4   4     5    7          55     E   NL  Hi     Age.DPO    35
## 5   5     5   10          58     L    L  Hi     Age.DPO    36
## 6   6     5    4          52     L    C  Hi     Age.DPO    36
```

```
tail(RxP.clean.long)
```

```
##         Ind Block Tank Tank.Unique Hatch Pred Res Measurement
## 24925 2497     3    5          29     L   NL  Lo log.Age.DPO
## 24926 2498     5   11          59     E    C  Lo log.Age.DPO
## 24927 2499     3    7          31     L    C  Lo log.Age.DPO
## 24928 2500     3    7          31     L    C  Lo log.Age.DPO
## 24929 2501     3    5          29     L   NL  Lo log.Age.DPO
## 24930 2502     3    5          29     L   NL  Lo log.Age.DPO
##             Value
## 24925    5.252273
## 24926    5.267858
## 24927    5.303305
## 24928    5.303305
## 24929    5.308268
## 24930    5.308268
```

Wow, that was pretty cool. We made our dataset super-duper long (almost 25,000 rows long!). Now, we can group the object by the "*Measurement*" column and calculate the mean values for each measurement with **summarize()**.

```
RxP.clean.long %>%
  group_by(Measurement) %>%
  summarize(Mean = mean(Value))
```

```
## # A tibble: 10 x 2
##    Measurement            Mean
##    <chr>                  <dbl>
##  1 Age.DPO                68.8
##  2 Age.FromEmergence      34.8
##  3 log.Age.DPO            4.15
##  4 log.Age.FromEmergence  3.20
##  5 log.SVL.final          2.94
##  6 Mass.final             0.399
##  7 Resorb.days            4.32
##  8 SVL.final              19.0
##  9 SVL.initial            18.6
## 10 Tail.initial           5.66
```

Okay, we can clearly see how to calculate the means. You can easily imagine how we could also group by our treatments to get all the means for all our measurements for all our treatment groups at once.

```
RxP.clean.long %>%
  group_by(Measurement, Pred, Res, Hatch) %>%
  summarize(Mean = mean(Value))
```

```
## # A tibble: 120 x 5
## # Groups:   Measurement, Pred, Res [60]
##    Measurement Pred  Res   Hatch Mean
##    <chr>       <fct> <chr> <chr> <dbl>
##  1 Age.DPO     C     Hi    E     55.3
##  2 Age.DPO     C     Hi    L     61.8
##  3 Age.DPO     C     Lo    E     99.2
##  4 Age.DPO     C     Lo    L     87.9
##  5 Age.DPO     NL    Hi    E     58.4
##  6 Age.DPO     NL    Hi    L     56.8
##  7 Age.DPO     NL    Lo    E     79.0
##  8 Age.DPO     NL    Lo    L     96.4
##  9 Age.DPO     L     Hi    E     48.9
## 10 Age.DPO     L     Hi    L     47.5
## # ... with 110 more rows
```

Now, let's see how we can spread that back out again to put the summarized data in a wide format. All we have to do is tack on one more line where we pipe the gathered and summarized object to a **spread()** function, and we specify which column will become the column titles and which column will be the data in those columns. These are once again called the "*key*," and the "*value*."

```
RxP.clean.long %>%
  group_by(Measurement, Pred, Res, Hatch) %>%
  summarize(Mean = mean(Value)) %>%
  spread(key=Measurement, value=Mean)
```

```
## # A tibble: 12 x 13
## # Groups:   Pred, Res [6]
##     Pred  Res   Hatch Age.DPO Age.FromEmergen~ log.Age.DPO
##     <fct> <chr> <chr>   <dbl>            <dbl>       <dbl>
##  1 C     Hi    E        55.3             21.3        3.98
##  2 C     Hi    L        61.8             27.8        4.09
##  3 C     Lo    E        99.2             65.2        4.51
##  4 C     Lo    L        87.9             53.9        4.42
##  5 NL    Hi    E        58.4             24.4        4.04
##  6 NL    Hi    L        56.8             22.8        4.02
##  7 NL    Lo    E        79.0             45.0        4.30
##  8 NL    Lo    L        96.4             62.4        4.47
##  9 L     Hi    E        48.9             14.9        3.87
## 10 L     Hi    L        47.5             13.5        3.85
## 11 L     Lo    E        59.5             25.5        4.02
## 12 L     Lo    L        61.4             27.4        4.08
## # ~ with 6 more variables: log.Age.FromEmergence <dbl>,
## #   log.SVL.final <dbl>, Mass.final <dbl>, Resorb.days <dbl>,
## #   SVL.final <dbl>, SVL.initial <dbl>, Tail.initial <dbl>
```

Pretty cool, right? The main point I want to make here is that *dplyr* and the other associated packages in the *tidyverse* make it really amazingly easy to manipulate your data and squish it this way and that, calculating different types of values. As you will see at the end of this chapter, this can be extremely useful when combined with plotting. It may seem somewhat daunting at first, but it is worth it to learn because it will make your life so much easier in the end.

9.4 EVEN MORE ADVANCED DATA WRANGLING! USING THE **DO()** FUNCTION

Being able to transform your data from long to wide formats has lots of advantages. One thing that is super handy is the ability to perform not just summary calculations on your data, but to do actual statistics in a tidy framework. For example, what if we wanted to know if the effect of predators in our experiment was consistent across blocks? You could code by hand eight different linear models, one for each block. But why not make our lives easier and automate it with *dplyr*?

First, we will group our data frame by "*Block*," then we will use the **do()** function to execute a simple linear model with "*Pred*" as our predictor. Just like using **summarize()**, whatever we put in the **do()** function will be done separately for each group that we specify (i.e., each block). There are a couple of other important points to be made here. First, we have to give a column name where the output of **do()** will be stored. Since we are fitting models, let's just call it "*mod.fit*." Second, even though we are piping the data frame to **do()** we still have to declare a data object within our linear model, and we do this by just calling the data frame a period, as in "*data=*." This tells the function to get the data for running the model from the pipe. Lastly, since we are making linear models here, we will use the tank mean data (*RxP.byTank*) so that we are not analyzing pseudo-replicated data. Okay, let's run the following code and examine the output.

```
#Here, we will run a lm on each block of the experiment
RxP.byTank %>%
  group_by(Block) %>%
  do(mod.fit = lm(log.SVL.final ~ Pred, data=.))
```

```
## # A tibble: 8 x 2
## # Rowwise:
##    Block mod.fit
```

```
##     <int> <list>
## 1      1 <lm>
## 2      2 <lm>
## 3      3 <lm>
## 4      4 <lm>
## 5      5 <lm>
## 6      6 <lm>
## 7      7 <lm>
## 8      8 <lm>
```

Hmm, that seems to have worked, but it didn't give us anything useful. Each object in our output column fit is a list, i.e., each one is the full model output. Here is where the **broom** package comes in handy. **broom** contains a function called **tidy()** which takes complex objects like model outputs and summaries and makes them into data frames, which makes them more compatible with lots of other functions. If we pipe the previous output to a line which runs the *mod.fit* column through the **tidy()** function, we can clean up our output a lot.

```
#Here, we will run a lm on each block of the experiment,
#and put the model outputs in a tidy format. Yay!
RxP.byTank %>%
  group_by(Block) %>%
  do(mod.fit = lm(log.SVL.final ~ Pred, data=.)) %>%
  tidy(mod.fit)
```

```
## # A tibble: 20 x 6
## # Groups:   Block [8]
##    Block term         estimate std.error statistic  p.value
##    <int> <chr>           <dbl>     <dbl>     <dbl>    <dbl>
## 1      1 (Intercept)     2.94     0.0212    139.    2.67e-16
## 2      1 PredNL          0.0346   0.0300      1.15  2.78e- 1
## 3      1 PredL           0.0835   0.0300      2.78  2.13e- 2
## 4      2 (Intercept)     2.93     0.0272    108.    2.56e-15
## 5      2 PredNL          0.00241  0.0384      0.0627 9.51e- 1
## 6      2 PredL           0.0822   0.0384      2.14  6.12e- 2
## 7      3 (Intercept)     2.95     0.0265    111.    1.95e-15
## 8      3 PredNL          0.0209   0.0375      0.556 5.92e- 1
```

```
##  9    3 PredL         0.125     0.0375    3.34    8.69e- 3
## 10    4 (Intercept)   2.91      0.0315   92.6     1.07e-10
## 11    4 PredL         0.101     0.0445    2.26    6.44e- 2
## 12    5 (Intercept)   2.89      0.0169  171.      6.06e-14
## 13    5 PredNL       -0.0325    0.0292   -1.11    3.02e- 1
## 14    5 PredL         0.0538    0.0239    2.26    5.85e- 2
## 15    6 (Intercept)   2.90      0.0128  227.      4.99e-13
## 16    6 PredL         0.0859    0.0181    4.74    3.19e- 3
## 17    7 (Intercept)   2.87      0.0299   95.7     8.78e-11
## 18    7 PredL         0.128     0.0424    3.03    2.30e- 2
## 19    8 (Intercept)   2.92      0.0293   99.5     6.94e-11
## 20    8 PredL         0.144     0.0415    3.48    1.31e- 2
```

Okay, now we are getting somewhere! However, this still isn't ideal. This is essentially the "*Coefficients*" section from the **summary()** output of eight different models, all just stacked on top of one another. It does give us the important details from the output for each block—we have the beta-coefficients for each predictor listed in alphabetical order and the associated test statistic and *P*-value as reported in the summary output. However, hopefully you remember that these *P*-values are not helpful. What we want is the *P*-value from the **Anova()** output, which is the *P*-value for the effect of "*Pred*" as a whole, not each level within it. Recall that the **Anova()** function (with the capital "A") is found in the *car* package. We can simply nest our linear model inside the **Anova()** function to get the proper model output.

Let's take a brief moment to reiterate what is going on here, because it really is kind of a lot. First, we are grouping our data frame by "*Block*," and thus anything we do after that will be done separately to each block in succession. Then, we are running a linear model, but that model is sitting inside the function **Anova()**, which is the exact same thing as storing the model as an object and then running **Anova()** on the object like we did in Chapters 5 and 6.

```
#Run a lm on each block of the experiment,
#but just save the Anova output
library(car)
```

```
RxP.byTank %>%
  group_by(Block) %>%
  do(mod.fit = Anova(lm(log.SVL.final ~ Pred, data=.))) %>%
  tidy(mod.fit)
```

```
## # A tibble: 16 x 6
## # Groups:    Block [8]
##     Block term         sumsq     df statistic  p.value
##     <int> <chr>        <dbl>  <dbl>     <dbl>    <dbl>
## 1       1 Pred        0.0141      2      3.91   0.0598
## 2       1 Residuals   0.0162      9      NA     NA
## 3       2 Pred        0.0175      2      2.96   0.103
## 4       2 Residuals   0.0266      9      NA     NA
## 5       3 Pred        0.0360      2      6.40   0.0187
## 6       3 Residuals   0.0253      9      NA     NA
## 7       4 Pred        0.0203      1      5.11   0.0644
## 8       4 Residuals   0.0238      6      NA     NA
## 9       5 Pred        0.0115      2      5.03   0.0442
## 10      5 Residuals   0.00796     7      NA     NA
## 11      6 Pred        0.0148      1     22.5    0.00319
## 12      6 Residuals   0.00394     6      NA     NA
## 13      7 Pred        0.0330      1      9.19   0.0230
## 14      7 Residuals   0.0215      6      NA     NA
## 15      8 Pred        0.0417      1     12.1    0.0131
## 16      8 Residuals   0.0206      6      NA     NA
```

Now we're talking! We have a simple output that shows us the *P*-values, *F*-statistics, and degrees of freedom for the linear models run on each block. Huzzah! We can see that predators significantly altered the final snout-vent length (SVL) of metamorphs in each block except Block 2.

Now that we've seen how to use the **do()** function, let's combine it with the **gather()** function to test for effects of all three treatment factors (Pred, Res, and Hatch) on every one of our response variables at once! Just as in the previous example, we will first gather our data into a very long format where our column headings are now listed in a column called "*Measurement*" and our actual data are in a column called "*Value.*" Then, we will group the data based on the "*Measurement*" column. Lastly, we will use **do()** to execute the linear model, using "*Value*" as our response variable and our three treatment variables as our predictors. This time let's assign the output to an object.

```
#Whoa, this does some amazing stuff
RxP.models <- RxP.byTank %>%
  gather(key = Measurement, value = Value,
         -Block, -Tank.Unique, -Hatch, -Pred, -Res) %>%
  group_by(Measurement) %>%
  do(mod.fit = Anova(lm(Value ~ Pred*Res*Hatch, data=.))) %>%
  tidy(mod.fit)
RxP.models
```

```
## # A tibble: 96 x 6
## # Groups:   Measurement [12]
##    Measurement        term        sumsq   df statistic   p.value
##    <chr>              <chr>       <dbl> <dbl>   <dbl>      <dbl>
## 1 Age.DPO             Pred        9.48e3    2   15.2     3.62e-6
## 2 Age.DPO             Res         9.70e3    1   31.2     4.75e-7
## 3 Age.DPO             Hatch       8.36e1    1    0.269   6.06e-1
## 4 Age.DPO             Pred:Res    2.57e3    2    4.13    2.04e-2
## 5 Age.DPO             Pred:Hatch  2.72e2    2    0.438   6.47e-1
## 6 Age.DPO             Res:Hatch   2.94e1    1    0.0945  7.60e-1
## 7 Age.DPO             Pred:Res:Ha... 5.25e2 2    0.845   4.34e-1
##  8 Age.DPO            Residuals   2.05e4   66   NA          NA
## 9 Age.FromEmergen... Pred        9.48e3    2   15.2     3.62e-6
## 10 Age.FromEmergen... Res        9.70e3    1   31.2     4.75e-7
## # ... with 86 more rows
```

And voila, just like that you have the full model output for all of the variables in the dataset! Wasn't that amazing?!?! I hope you agree that it was. Of course, this only works in cases where you are running the same basic model on different variables. But we can also do something really simple to filter this output based on the "*p.value*" column to see which predictors have significant effects on which variables.

```
#Just show the predictors with a p-value below 0.05
RxP.models %>%
  filter(p.value < 0.05)
```

```
## # A tibble: 20 x 6
## # Groups:    Measurement [12]
##    Measurement      term      sumsq    df statistic   p.value
##    <chr>            <chr>     <dbl> <dbl>     <dbl>     <dbl>
##  1 Age.DPO          Pred     9.48e+3     2      15.2 3.62e- 6
##  2 Age.DPO          Res      9.70e+3     1      31.2 4.75e- 7
##  3 Age.DPO          Pred:R~  2.57e+3     2      4.13 2.04e- 2
##  4 Age.FromEmergen~ Pred     9.48e+3     2      15.2 3.62e- 6
##  5 Age.FromEmergen~ Res      9.70e+3     1      31.2 4.75e- 7
##  6 Age.FromEmergen~ Pred:R~  2.57e+3     2      4.13 2.04e- 2
##  7 log.Age.DPO      Pred     1.95e+0     2      17.8 6.60e- 7
##  8 log.Age.DPO      Res      1.78e+0     1      32.6 2.97e- 7
##  9 log.Age.FromEme~ Pred     9.71e+0     2      16.6 1.40e- 6
## 10 log.Age.FromEme~ Res      8.34e+0     1      28.6 1.20e- 6
## 11 log.SVL.final    Pred     1.65e-1     2      26.2 4.20e- 9
## 12 Mass.final       Pred     3.29e-1     2      20.4 1.30e- 7
## 13 N.alive          Pred     6.03e+3     2      32.7 1.35e-10
## 14 N.alive          Hatch    4.62e+2     1      5.02 2.84e- 2
## 15 N.dead           Pred     6.03e+3     2      32.7 1.35e-10
## 16 N.dead           Hatch    4.62e+2     1      5.02 2.84e- 2
## 17 Resorb.days      Res      4.37e+0     1      14.8 2.78e- 4
## 18 SVL.final        Pred     6.45e+1     2      25.6 6.06e- 9
## 19 SVL.initial      Pred     5.55e+1     2      24.5 1.12e- 8
## 20 Tail.initial     Pred     9.22e+0     2      7.36 1.30e- 3
```

This quickly reveals (as we probably know by now) that predators have the strongest effects across the board, and that they occasionally interact with resource level. Pretty cool right? In reality we should not be running linear models on all of these variables, since several of them are non-normal (as we saw in Chapter 7) and **lm()** is not appropriate for analyzing them.

Let's work through one more example before we are done. What about more complex models like the linear mixed effects models we explored in

Chapter 8? We can still use **do()** to make our lives easier! You have two options here.

1. You can use **Anova()** from the *car* package to obtain summary statistics for your model. Recall however that for mixed models you should use nested likelihood ratio tests (LRTs) to obtain the most accurate estimates of significance. That said, using **Anova()** in this case is a very effective method to quickly get a glance at the *likely* significance of your predictors.
2. You can run a likelihood ratio test inside of **do()**, but only for a single predictor at a time. This is the same basic idea as nesting the function inside of the **Anova()** function, but now we are nesting *two* functions together and using *anova()* to calculate a likelihood ratio test on them.

The code for both examples is shown in the following section, but I'll only show the output for the first option.

```
library(lme4)
#Get quick and dirty estimates of p-values
RxP.clean %>%
  gather(key = Measurement, value = Value,
         -Ind, -Block, -Tank, -Tank.Unique,
         -Hatch, -Pred, -Res) %>%
  group_by(Measurement) %>%
  do(mod.fit = Anova(lmer(Value ~ Pred*Res*Hatch+
                          (1|Block/Tank), data=.))) %>%
  tidy(mod.fit)
```

```
## # A tibble: 70 x 5
## # Groups:   Measurement [10]
##    Measurement      term        statistic   df    p.value
##    <chr>            <chr>           <dbl> <dbl>      <dbl>
## 1 Age.DPO          Pred             32.6     2    8.22e-8
## 2 Age.DPO          Res              36.6     1    1.43e-9
## 3 Age.DPO          Hatch            0.261    1    6.10e-1
## 4 Age.DPO          Pred:Res         8.85     2    1.20e-2
```

```
## 5 Age.DPO          Pred:Hatch    0.991    2    6.09e-1
## 6 Age.DPO          Res:Hatch     0.102    1    7.49e-1
## 7 Age.DPO          Pred:Res:H... 2.20     2    3.33e-1
## 8 Age.FromEmergen~ Pred          32.6     2    8.22e-8
## 9 Age.FromEmergen~ Res           36.6     1    1.43e-9
## 10 Age.FromEmergen~ Hatch        0.261    1    6.10e-1
## # ~ with 60 more rows
```

```
#Get an accurate p-value for a single predictor at a time
#Here, we are evaluating the interaction between Pred and Res
#on each of our response variables
RxP.clean %>%
  gather(key = Measurement, value = Value,
         -Ind, -Block, -Tank, -Tank.Unique, -Hatch, -Pred, -Res)
  %>%
  group_by(Measurement) %>%
  do(mod.fit = anova(lmer(Value ~ Pred*Res+(1|Block/Tank), data=.),
                     lmer(Value ~ Pred+Res+(1|Block/Tank), data=.))) %>%
  tidy(mod.fit)
```

9.5 MAKING BETTER FIGURES WITH *ggplot2*

Box 9.4 - Principles of effective figure making

Before we get into the meat of how to most efficiently use **ggplot2** for plotting, it is useful to take a moment to talk about what makes a good figure. What is it that makes a nice looking figure, one that is suitable for publication or use in a presentation? I would argue that 1) judicious use of color, 2) large clear text and labels, and 3) efficient usage of plot space are three hallmarks of a good figure. It is often useful to create multiple panels to show different aspects of your data. These fundamentals are the same whether you are making your figures in R or not.

There are two main ways to make figures in R: base graphics (i.e., those that are built in to the base version of R you downloaded from CRAN) and using the package **ggplot2**. There are functions in other packages (e.g., the **scatterplot()** function in the *car* package, or the **barplot2()** function in the *gplots* package) but these all utilize the coding of base graphics. While **ggplot2** and base graphics use different coding styles, the fundamentals that make an effective graphic remain the same. Both types of coding allow you to build your graphics piece by piece and really give you control over every aspect of the figure. We will just focus on **ggplot2** here, as it is vastly superior to base graphics.

Let's talk about some basic nuts and bolts that are useful to know.

9.5.1 Defining colors

As with so many things in R, there are multiple ways to define what colors you want to use. The various ways to define colors were described in detail in Chapter 4, so here is just a super-quick recap. You can define colors by number, by name, by red, green, and blue (rgb), or by the hexidecimal system. Each method has it pros and cons. Here, we will use named colors.

9.6 BASICS OF *ggplot2*

In recent years, the **ggplot2** package has become increasingly popular for making nice looking figures with relative ease. In particular, **ggplot2** is excellent for exploratory data analysis, as it easily allows you to plot (for example) histograms for each of your groups or side-by-side. As we've already seen, you can easily add linear regression lines and confidence intervals to a scatterplot. But **ggplot2** can be used to make pretty much any type of graphic you want, you just have to know how (which is, of course, the same dilemma in base graphics!). This book cannot attempt to teach you everything you need to know about using **ggplot2**, but is instead meant to give you an introduction and provide you with tools to get started. As mentioned previously, there are whole books just on using **ggplot2**!

The downside of **ggplot2** is that it uses a very different syntax from base graphics. The basic ideas are the same, in that you layer objects together to create a final product, but the way you get there is quite different. Like the **tidyverse**, **ggplot2** is written by Hadley Wickham, which is in part why **ggplot2** and packages like **dplyr** work so well together!

It is worth pointing out that the folks at RStudio have made some very handy cheatsheets for using these packages. I highly recommend checking them out. A quick internet search for "ggplot2 cheatsheet" will help you find them quickly.

Box 9.5 - Understanding the grammar of *ggplot2*

There are two basic functions that do all the heavy lifting for plotting: **qplot()**, which stands for "quick plot" and **ggplot()**. **qplot()** is great for exploratory data analysis, as it makes intuitive nice looking figures with very little code. However, if you want to make truly professional and custom figures, you probably want to use **ggplot()**. In both cases, you have the following basic elements:

- You assign the data you are going to use up front, using the "*data=*" argument.
- You assign aesthetics such as the colors you want to use, what data will be on the x-axis or y-axis, etc., using the **aes()** argument. In **qplot()** there is no actual **aes()** argument, but the principle is the same.
- You assign the basic style of the graphic (e.g., scatterplot, histogram, etc.) using the **geom()** argument, which defines the geometric interpretation of the data.

You can also use the following arguments to further customize your graphic:

- You can divide your plot into multiple panels, or *facets*.
- You can change the automatically generated coordinate system with *coord*.
- You can change aspects of the axes or size of the plotted objects with *scales*.
- You can use *themes* to apply a set of defaults which will affect how your plot looks. As is probably apparent by now, I quite like **theme_cowplot**.

A really cool feature of ***ggplot2*** is that you can assign your plot to an object, just like it was a vector or a data frame. The object retains a copy of the data used to make the plot, and all the code used to make the plot. You can then add features to the existing plot. This is also useful for plotting different figures together using the **plot_grid()** function in the ***cowplot*** package, which I will do next.

9.6.1 Getting started with **qplot()**

We will start by walking through a series of increasingly complex plots made in **qplot()** and then explore the various ways you customize your plot (which is the same if you use **qplot()** or **ggplot()**). Some of this will repeat material covered in earlier chapters, but I've put it here to be complete.

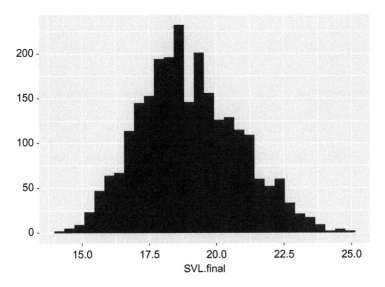

Figure 9.1 A very boring histogram of metamorph SVL's.

Let's start by making a simple histogram (Figure 9.1). Recall that for any plot we have to define where the data are found, our aesthetics (x- and y-variables, colors, etc.), and the geometric interpretation of the data.

```
qplot(data=RxP.clean,
      x=SVL.final,
      geom="histogram")
```

Notice the setup of the previous code. Instead of using a function to specify the specific type of plot we want to make (e.g., **hist()**) we use a more generic plotting function (**qplot()**) and then specify the type of plot we want to make with the "*geom=*" argument. In addition, we just list the column we want to plot and provide the data frame where the data can be found. In truth, if you do not provide a geom to use, **qplot()** will guess what you want to use (and histograms are the default for continuous numerical data). As you have probably noticed by now, the default plot in **ggplot2** has a light grey grid behind it. If you love that look, great! But, it is easy to change the theme, and in particular you can change the default theme for an entire R session. Here, let's change the default theme to **theme_cowplot()**.

```
library(cowplot)
```

```
## 
## ***********************************************************
## Note: As of version 1.0.0, cowplot does not change the
##    default ggplot2 theme anymore. To recover the previous
##    behavior, execute:
##    theme_set(theme_cowplot())
## 
## ***********************************************************
```

```
theme_set(theme_cowplot())
```

The previous histogram does not make use of any of ***ggplot2***s unique capabilities. For example, what if we want to examine histograms for each of our predator treatments? Here are two examples of how you might compare the three treatments by plotting the histograms together. Note that all we have to do is specify how we want the geom to be colored or filled and **qplot()** knows what to do. The first example changes the color of the outside of the bars, whereas the second example changes the fill of each bar. The third and fourth examples are not histograms but instead are density plots, which also give you an idea of the relative spread of data in each treatment (Figure 9.2).

```
#Make 4 plots to explore the difference between
#filling and coloring histograms and density plots
a<-qplot(data=RxP.clean,
         x=SVL.final,
         geom="histogram",
         col=Pred)
b<-qplot(data=RxP.clean,
         x=SVL.final,
         geom="histogram",
         fill=Pred)
c<-qplot(data=RxP.clean,
```

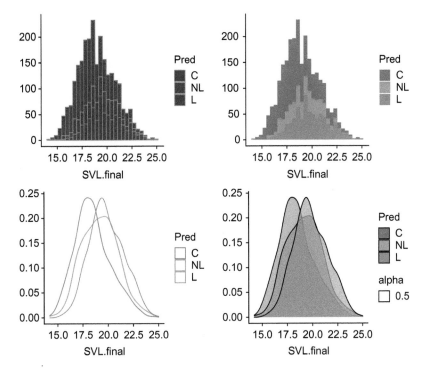

Figure 9.2 Slightly more interesting figures.

```
         x=SVL.final,
         geom="density",
         col=Pred)
d<-qplot(data=RxP.clean,
         x=SVL.final,
         geom="density",
         fill=Pred,
         alpha=0.5) #this one is translucent!
plot_grid(a,b,c,d, nrow=2)
```

Notice what happened in those figures. When we specified to *color* the histogram or density plot it changed the outline color whereas when we changed the *fill* color, it filled in the bars or density plot. Lastly, in the fourth plot we specified an alpha level of 0.5, which made our density plots translucent.

Yet another way to view these histograms would be to use ***ggplot2***'s faceting feature, which easily allows us to split a single figure into multiple panels to allow easy visual comparison. For example, we can use the "*facets=*" argument to specify how we want our figure split apart. Facets are coded in terms of "*rows~columns*." The most important thing to know about specifying facets is that you *must* specify which data should be plotted in *both* the rows and columns, with a "~" in between them (Figure 9.3). If we want to examine histograms for each combination of predation and resource treatment, we can do so by typing the following:

```
qplot(data=RxP.clean,
      x=SVL.final,
      geom="histogram",
      facets=Res~Pred, #facet the plots into panels
      fill=Pred)
```

One great thing about the faceting feature is that all panels are placed on the exact same axes, so that they are easily comparable as you look across them. In the previous example, we plotted rows (the left-hand side of the ~)

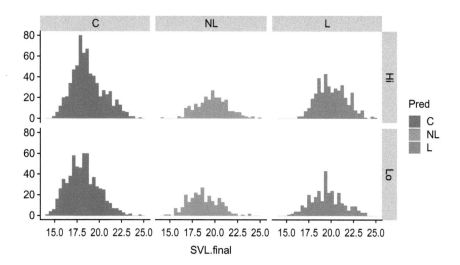

Figure 9.3 Data on snout-vent length (SVL) have been plotted for each combination of predator and resource treatment and then filled based on predator treatment.

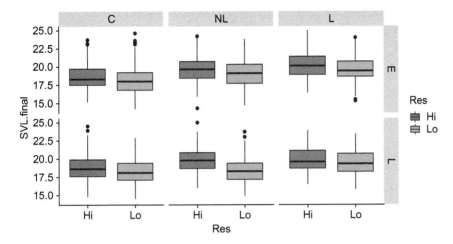

Figure 9.4 Data on SVL have been plotted for each resource treatment, and then faceted by predator and hatching treatments.

based on resource treatment, and columns (the right-hand side of the ~) based on predation treatment. If you only want to plot the predator treatments next to one another (for example), we would have placed a period (.) on the side we wanted blank, as in "*facets=.~Pred.*"

We can make a large number of different plots using just the basic **qplot()** function. For example, we can make boxplots, violin plots, scatterplots, and so on. Perhaps we want to quickly look at the box and whisker plots of all of our 12 treatment combinations. This is easily accomplished by plotting SVL against one treatment, then faceting based on the other two (see Figure 9.4).

```
qplot(data=RxP.clean,
      y=SVL.final,
      x=Res,
      geom="boxplot",
      facets=Hatch~Pred, #facet in two dimensions
      fill=Res)
```

All of this coding works equally well for scatterplots, where we have two continuous variables plotted against one another. For example, maybe we want to plot the relationship between "*SVL.final*" and "*Mass.final*," and we want to visualize it across predator and resource treatments. Here, we will

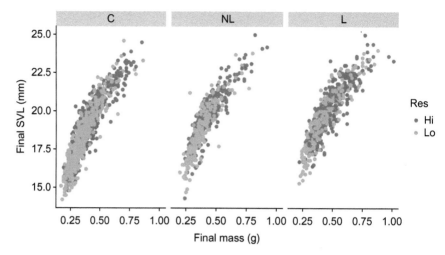

Figure 9.5 Data on SVL plotted against final mass at metamorphosis with points colored for each resource treatment, and faceted by predator treatment.

change the geom to "*point.*" Notice that we also have to *color* the points. You can't *fill* points. Also, notice in the following code we've added more meaningful x- and y-axis titles by adding "*xlab=*" and "*ylab=*" arguments (Figure 9.5). These don't work in **ggplot()**, just in **qplot()**. You can of course change the axis titles in **ggplot()**, it is just through a slightly different mechanism.

```
qplot(data=RxP.clean,
      y=SVL.final,
      x=Mass.final,
      geom="point",
      facets=.~Pred, #facet by Pred treatment
      col=Res,
      xlab="Final mass (g)",
      ylab="Final SVL (mm)")
```

We can even make a very cool variation of a scatterplot commonly called a "bubblechart." This is accomplished by assigning the size of points to be based on some continuous variable in our dataset. For example, in Chapter 7, we created a variable in our tank summarized dataset (*RxP.byTank*) called "*N.dead*" which was the number of tadpoles that died in each mesocosm. We can easily specify that the size of points should be determined by the values in our "*N.dead*" column. We can also make our points partially translucent by specifying an "*alpha*" level. Here, let's plot the size of metamorphs against the age at which they crawled out of the water, while plotting the number of individuals that died during the larval period as the size of each point (Figure 9.6).

```
qplot(data=RxP.byTank,
    y=SVL.final,
    x=Age.DPO,
    size=N.dead, #specify the size of points
    col=Res,
    alpha=0.5,
    xlab="Final mass (g)",
    ylab="Final SVL (mm)")
```

Lastly, we can even add a linear regression with confidence intervals to a scatterplot made with **qplot()**. All we do is make a scatterplot and then add to it a second geom argument to make a "smooth" line, which is the regression fit with confidence intervals (Figure 9.7). You would still want

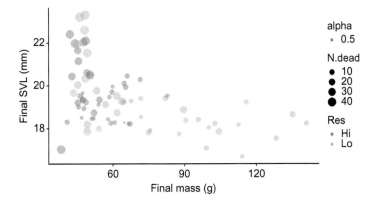

Figure 9.6 Data on SVL have been plotted against age at metamorphosis with points colored for each resource treatment, and sized based on the number of animals that died in each tank. A plot like this quickly demonstrates that the largest metamorphs also came from tanks with the highest mortality, and most of those animals also metamorphosed earliest.

to run the linear model and inspect the **summary()** output as you do not receive any actual statistical information from **qplot()** (i.e., this is just for visualization). Note that the default is a smoothed curve and not a linear regression, so to add a regression fit we specify "*method=lm*." Also, note that while this is very easy to do for an lm it is a little more complicated for a glm or other non-linear model, as we saw in Chapters 6 and 7.

```
qplot(data=RxP.byTank,
      x=Age.DPO,
      y=SVL.final,
      geom="point",
      xlab="Age at metamorphosis (d)",
      ylab="Final SVL (mm)")+
  geom_smooth(method="lm")#it is so easy to add a regression
```

We can even combine the faceting we did earlier for simple figures like histograms with the model fitting that is occurring when we add in a regression line. The key here is that once we specify something like facets or colors, any commands that follow will assume those rules. Although Figure 9.8 looks very complicated, the principles are exactly the same. We've just specified (in this order) where the data are, the x-axis variable, the y-axis variable, the color to split the data by, the geom, the variable to

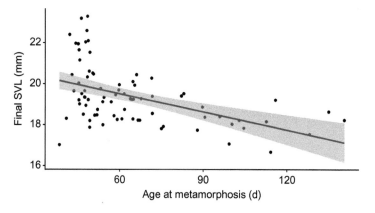

Figure 9.7 A linear regression with confidence interval is shown plotted on the scatterplot of *SVL.final* against *Age.DPO*. Plotting regressions is so easy with ***ggplot2***!

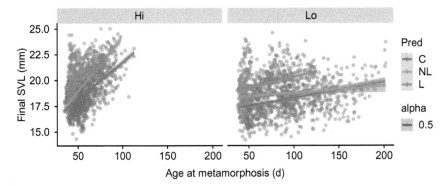

Figure 9.8 Linear regressions with confidence intervals are shown for each predator and resource treatment combination, plotting age at metamorphosis against SVL at metamorphosis. Wow, that was amazingly easy!

facet the data by, that we want the points to be translucent, axis titles, and that we want to add linear regressions (Figure 9.8).

```
qplot(data=RxP.clean,
      x=Age.DPO,
      y=SVL.final,
      col=Pred,
      geom="point",
      facets=.~Res,
      alpha=0.5,
      xlab="Age at metamorphosis (d)",
      ylab="Final SVL (mm)")+
   geom_smooth(method="lm")
```

9.7 CUSTOMIZING YOUR FIGURE

There are many, many, many different ways to customize your figure. Here is a decidedly non-exhaustive list of important and useful functions to know.

9.7.1 Changing colors

While the default colors chosen for you in *ggplot2* are pretty appealing, sometimes you want to specify the colors *you want*. When you change the ways that the data are mapped to the figure, you are changing a *scale*. Thus, to customize your colors, you are usually using either **scale_fill_manual()**

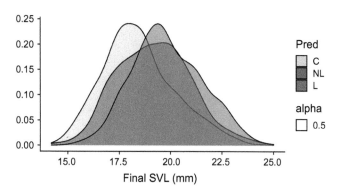

Figure 9.9 A density plot of the final SVL at metamorphosis, with fun colors of our choosing!

or **scale_color_manual()**, depending on if you have filled your object or colored it. For example, in our density plot earlier, we filled based on our predator treatment. Let's override the colors using **scale_fill_manual()**. We assign the new colors of our choice with the argument "*values=*" and the new colors have to be concatenated together in a vector (Figure 9.9).

```
qplot(data=RxP.clean,
      x=SVL.final,
      geom="density",
      fill=Pred,
      alpha=0.5,
      xlab="Final SVL (mm)")+
  scale_fill_manual(values=c("sky blue", #Oooh, now we
                             "seagreen", #are adding
                             "orchid"))  #fun colors!
```

Using **scale_fill_manual()** or **scale_color_manual()** also allows us to modify the text in our legend. We can modify the way that our treatment levels are displayed with the "*labels=*" argument, and we can change the name of the legend with the "*name=*" argument (Figure 9.10).

```
qplot(data=RxP.clean,
      x=SVL.final,
      geom="density",
      fill=Pred,
      alpha=0.5,
```

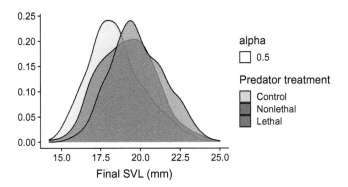

Figure 9.10 A density plot with custom colors, and a more meaningful legend.

```
    xlab="Final SVL (mm)")+
scale_fill_manual(values=c("sky blue",
                           "seagreen",
                           "orchid"),
               name="Predator treatment",
               labels=c("Control",   #now we can change
                        "Nonlethal",#the way the legend
                        "Lethal"))   #is displayed
```

9.7.2 Changing the axes limits

Sometimes you will want to change the limits of your x- or y-axes from whatever *ggplot2* decides to use. Generally, the choices that are made are very sensible, but occasionally you will want to set them to what you want. We will see an example of this later with a bar graph, but for now, let's imagine you want to plot the scatterplot we made earlier showing the values all the way to the origin (I don't know why you would want to, but let's just do it as an example). You can change the limits of your axes by adding **coord_cartesian()** to your plot, and within the parentheses you specify your x- and y-axes with "*xlim=*" and "*ylim=*" arguments. For each argument, you concatenate together the min and max values of the axis into a two-number vector (Figure 9.11).

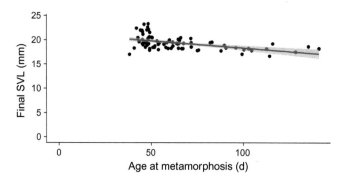

Figure 9.11 We changed the axes for some reason.

```
qplot(data=RxP.byTank,
      x=Age.DPO,
      y=SVL.final,
      geom="point",
      xlab="Age at metamorphosis (d)",
      ylab="Final SVL (mm)")+
  geom_smooth(method="lm")+
  coord_cartesian(xlim=c(0,140),  #change the x- and
                  ylim=c(0,24))   #y-axis limits
```

9.7.3 A little more about themes

Before this chapter, nearly all of the figures we've made have used the default ***ggplot2*** theme, which has the light grey background with faint gridlines. In this chapter, I've changed the default to **theme_cowplot()**. There are many different built-in themes. Themes allow you to customize nearly every little thing, such as the font size of your axis text. You can even make your own theme and save it! To give you an example of how the choice of theme affects your plot, I've created three plots in the following section, each using a different theme.

In the next code, notice that I have assigned the first plot to be object "*a*," then have made objects "*b*" and "*c*" by adding new elements to "*a*." Recall from earlier that when you assign your ***ggplot2*** figure to an object, it retains everything needed to make the plot. It's not just a visual thing, it contains all the data needed to make the figure. When you add new code to

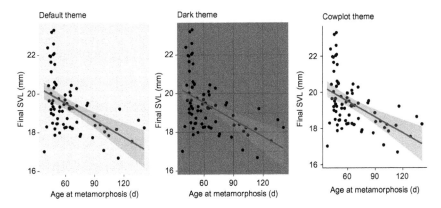

Figure 9.12 Three different themes show how defining a theme changes the way your figure looks.

the plot, for example a theme or a title as we've done in the next example, it overrides the existing version of that code. So in plot "*a*," I've added a title with the **ggtitle()** function, but then we write over that in plots "*b*" and "*c*" (Figure 9.12).

```
a<-qplot(data=RxP.byTank,
         x=Age.DPO,
         y=SVL.final,
         geom="point",
         xlab="Age at metamorphosis (d)",
         ylab="Final SVL (mm)")+
   geom_smooth(method="lm")+
   theme_grey()+
   ggtitle(label="Default theme")
b<-a + theme_dark()+
   ggtitle(label="Dark theme")
c<-a + theme_cowplot()+
   ggtitle(label="Cowplot theme")
plot_grid(a,b,c, nrow=1)
```

9.7.4 Advanced faceting

Up until now, we have faceted our plots in **qplot()** using the "*facets=*" argument. This has two potential drawbacks. 1) Sometimes you have many panes you would like to create in a single dimension, so many that they would need to wrap around into multiple rows. 2) While it can be really nice to have your plots all be on the same axes, sometimes you don't want that. For example, Figure 9.8 has a lot of unused space in the left panel. There are two functions for faceting outside of what is built in to **qplot()**. The first is **facet_grid()**, which does essentially the same thing as what we did before in **qplot()**. The other is **facet_wrap()**, which allows you to wrap your facets into multiple rows. The advantage of both of these is that you can include the argument "*scales='free'*" which will let the axes vary independently for each panel. For example, we can modify our code from earlier when we made linear regressions faceted by resource treatment. Now, instead of the faceting being coded inside the **qplot()** function, we code it separately in **facet_grid()** (Figure 9.13). You'll see an example of using **facet_wrap()** a little later.

```
qplot(data=RxP.clean,
      x=Age.DPO,
      y=SVL.final,
      col=Pred,
      geom="point", alpha=0.5,
      xlab="Age at metamorphosis (d)",
      ylab="Final SVL (mm)")+
  geom_smooth(method="lm")+
  facet_grid(facets=.~Res, scales="free")
```

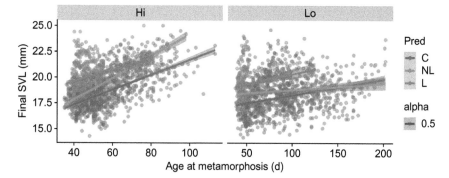

Figure 9.13 Notice now that the x-axes are allowed to be different from one another, since we set the scales to be "free."

9.7.5 Changing your facet labels

We've already seen how to modify the text in a legend. What about the text at the top of the facet headings? That is also done in the **facet_grid()** or **facet_wrap()** function and it uses another nested function called **labeller()**. In this function, you define new labels for each of the levels in your factor. For example, we have faceted based on the resource treatment, which has two levels, Hi and Lo. If we want to rename those in our plot, we can do so using the **labeller()** function. Somewhat confusingly, you implement this with an argument that is also called "*labeller=*" (Figure 9.14).

```
qplot(data=RxP.clean,
      x=Age.DPO,
      y=SVL.final,
      col=Pred,
      geom="point",
      alpha=.5,
      xlab="Age at metamorphosis (d)",
      ylab="Final SVL (mm)")+
  geom_smooth(method="lm")+
  facet_grid(facets=.~Res, scales="free",
             labeller=labeller(Res=c(Hi = "High resources",
                                      Lo = "Low resources")))
```

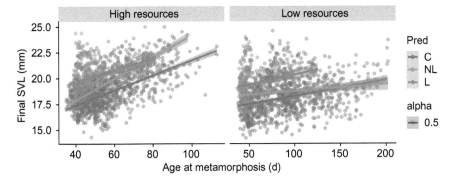

Figure 9.14 Now our facets labels are much more intuitive than just the short codes we used in our data file. Hooray!

9.7.6 Using **ggplot()** vs **qplot()**

Using the **ggplot()** function instead of **qplot()** has two major effects: 1) It is slightly more cumbersome to use because now you are responsible for coding more of the basic elements of the figure, but 2) it opens up the possibilities of the kinds of figures you can make, because now you have complete control. **qplot()** does some thinking for you. If you provide it with some data but don't specify a geom, it will make a histogram or a dotplot or a scatterplot by default. **ggplot()** won't do that, you now have to specify "**+geom_histogram()**" or "**+geom_boxplot()**," for example.

In some respects, using **ggplot()** becomes a process more similar to base graphics. Using **qplot()**, you are often interested in a quick look at your data and in exploring patterns. Thus, the default colors and symbols that are chosen for you are perfectly fine. And to be honest, the default parameters are often quite visually pleasing. However, when you want to specify exact colors or complex nonlinear regression fits (as just two examples), the process becomes more laborious. That said, once you get used to the language of *ggplot2*, it is likely that you will never want to use base graphics again.

9.8 COMBINING DATA WRANGLING WITH PLOTTING
       WITH *ggplot2*

Now that we've seen how to calculate our means and standard errors, let's explore how to directly pipe that to a plotting function. Remember, everything in the *tidyverse* is designed to work together. Earlier we made an object called RxP.summary that contained the means and standard errors for Age.DPO. Why not just pipe it directly to *ggplot2*? Since we are piping the summarized object, we have to keep in mind what will be the actual input to the plotting function. There is one more thing to keep in mind: even though we pipe the object into the **ggplot()** function, we still need to provide a "*data=*" argument. In order for **ggplot()** to know that the data object is what it is receiving via the pipe function, we just specify "*data=.*," just like we did before with the **do()** function.

9.8.1 Let's make a professional looking bargraph!

Lots of people these days do not like bargraphs, but they are still pretty common, so it is useful to go through how you make one in ***ggplot2***. Here is the full code to make Figure 9.15. Don't worry, we will walk through what all of that code means line by line. The first half of this code is exactly what we did earlier in the chapter to calculate the means and standard errors of our data, and the second half is the code for plotting.

```
RxP.clean %>%
  group_by(Pred, Res, Hatch) %>%
  summarize(Mean.Age.DPO = mean(Age.DPO),
            SD.Age.DPO = sd(Age.DPO),
            N.Age.DPO = length(Age.DPO)) %>%
  mutate(SE.Age.DPO = SD.Age.DPO/sqrt(N.Age.DPO)) %>%
  ggplot(data=., aes(x = Pred, y = Mean.Age.DPO, fill=Res)) +
    geom_col(position="dodge", col="black") +
    geom_errorbar(aes(ymax = Mean.Age.DPO + SE.Age.DPO,
                      ymin = Mean.Age.DPO - SE.Age.DPO,
                      width=0.4),
                  position=position_dodge(width=0.9)) +
    facet_grid(facets=.~Hatch)+
    ylab(label="Age at emergence (mean ± SE)")+
    xlab(label="Predator treatment")
```

Well, that worked really well! Let's unpack all of the plotting code, which includes six different functions.

```
ggplot(data=., aes(x = Pred,
                   y = Mean.Age.DPO,
                   fill=Res)) +
```

1) We use **ggplot()** to define the basic parameters of the figure. First, we define where the data will come from (in this case, the pipe), then assign the aesthetics which includes that our x-axis is "*Pred,*" our y-axis is "*Mean.Age.DPO,*" and that we want to fill in our data based on "*Res.*"

```
geom_col(position="dodge", col="black") +
```

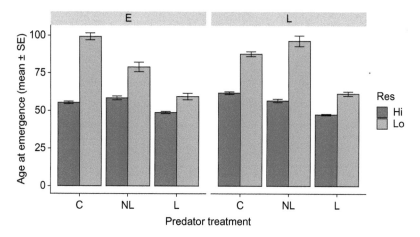

Figure 9.15 A really nice looking figure that shows the means ± standard errors of age at emergence of red eyed treefrog metamorphs, plotted for all 12 combinations of predators, resources, and hatching ages.

2) We assign a geometric interpretation (a "geom") for how **ggplot()** should transform and plot our data, which in this case is **geom_col()**. Although we often call these "barplots" or "bar graphs," in the **ggplot2** world, a *bar* plot has horizontal bars whereas *columns* are vertical. We also have to define that we want our filled bars next to one another instead of stacked (which is the default) which is done by "dodging" them, and we've defined that the border color of the bars is black.

```
geom_errorbar(aes(ymax = Mean.Age.DPO + SE.Age.DPO,
                  ymin = Mean.Age.DPO - SE.Age.DPO,
                  width=0.4),
              position=position_dodge(width=0.9)) +
```

3) We define the location and size of the error bars with another geom, **geom_errorbar()**. We define three aesthetics: the max and min of each bar (which is just the mean + or - the SE calculation) and the width of the bars (the default is as wide as each bar). Lastly, we define the position of the bars, which is the line

"position=position_dodge(width=0.9)." Why *"width=0.9"* you are asking? The default width of each bar is 0.9 (out of 1). If we had defined our bars as "width=0.5" in the **geom_col()** function, we would have then defined our errorbars to be "width=0.5" as well. Essentially, we are telling the error bars that they need to be plotted in the right location for bars with a width of 0.9.

```
facet_grid(facets=.~Hatch)+
```

4) We faceted our plot based on hatching age using the **facet_grid()** function. Recall from earlier that faceting requires a rows and columns setup, separated by the "~." Here, we said to make no rows (.) and facet the columns based on hatching age.

```
ylab(label="Age at emergence (mean ± SE)")+
xlab(label="Predator treatment")
```

5) We changed our y-axis label to make it more meaningful.
6) We changed our x-axis label to make it more meaningful. These are essentially the same lines of code you would use in base graphics.

There are a still a few things we might want change about this figure. For example, our treatment labels are the short codes that are present in our data frame, but they are not very helpful in our figure. How can we fix those? There are of course multiple ways one could do that, but one way that might be handy would be to recode the variables using **mutate()**, which would allow us to reassign the names of the factor labels using the *"labels="* argument. We can do this before we plot the function. Note that, we could also do this with the **labeller()** function that we've already explored, but this will show you yet another way to tackle the same problem! Another thing we might want to do is customize our colors using **scale_fill_manual()**, which also allows us to modify the name of our figure legend. One advantage of this type of coding is that we haven't modified our original data object (*RxP.clean*) at all, nor have we cluttered our workspace with lots of temporary objects?. All of the modifications we have made only exist in this chunk of code that is making the figure. Alternatively, you could easily assign the data object you produce here to be an object in the memory, then just refer to that in the "*data=*" argument of the **ggplot()** call (Figure 9.16).

```
RxP.clean %>%
  mutate(Pred = factor(Pred, labels=c("Control",
                                      "Nonlethal",
                                      "Lethal")),
         Hatch = factor(Hatch, labels=c("Early hatched",
                                        "Late hatched")),
         Res = factor(Res, labels=c("High resources",
                                    "Low resources"))) %>%
  group_by(Pred, Res, Hatch) %>%
  summarize(Mean.Age.DPO = mean(Age.DPO),
            SD.Age.DPO = sd(Age.DPO),
            N.Age.DPO = length(Age.DPO)) %>%
  mutate(SE.Age.DPO = SD.Age.DPO/sqrt(N.Age.DPO)) %>%
  ggplot(data=., aes(x = Pred,
                     y = Mean.Age.DPO,
                     fill=Res)) +
  geom_col(position="dodge", col="black") +
  geom_errorbar(aes(ymax = Mean.Age.DPO + SE.Age.DPO,
                    ymin = Mean.Age.DPO - SE.Age.DPO,
                    width=0.4),
                position=position_dodge(width=0.9)) +
  facet_grid(facets=.~Hatch)+
  ylab(label="Age at emergence (mean ± SE)")+
  xlab(label="Predator treatment")+
  scale_fill_manual(values=c("seagreen","skyblue"),
                    name="Resource treatment")
```

That probably seems like a *ton* of code, but if you think about the fact that we've been building it bit by bit, it really isn't that bad.

9.8.2 Now let's really do something cool: combining gathered data and faceting

Many times, it is useful to look at lots of your data at once. Perhaps you want to see if any variables show you any interesting trends. Or maybe you just don't want to have to make ten versions of the same figure. In

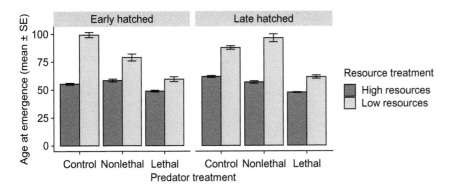

Figure 9.16 Now THAT is a good looking figure!

any event, let's combine what we learned earlier about gathering data from a wide to long format and what we know about faceting to make boxplots of all of our response variables at once. First, we will gather our *RxP.clean* data into a long format as before, where we have a column called "*Measurement*" with the former column headings and a column called "*Value*" with the values that used to be in each column. Then, we pipe that object to ggplot and define our aesthetics. Here, I want to make boxplots with the predator treatment on the x-axis and have boxes filled by resource treatment. After specifying the geom, we use **facet_wrap()** to facet based on the "*Measurement*" column and we specify that we want our scales to be free, which is particularly important since the different measurements are on *very* different scales (Figure 9.17).

```
RxP.clean %>%
  gather(key = Measurement, value = Value,
         -Ind, -Block, -Tank, -Tank.Unique,
         -Hatch, -Pred, -Res) %>%
  ggplot(data=., aes(x=Pred, y=Value, fill=Res))+
  geom_boxplot()+
  facet_wrap(facets=.~Measurement, scales="free", nrow=5)
```

Now, we will do one last thing that brings everything together. Since we know how to gather and summarize lots of data at once, and we know how to gather our data and pipe it to **ggplot()**, why not do both? In one single

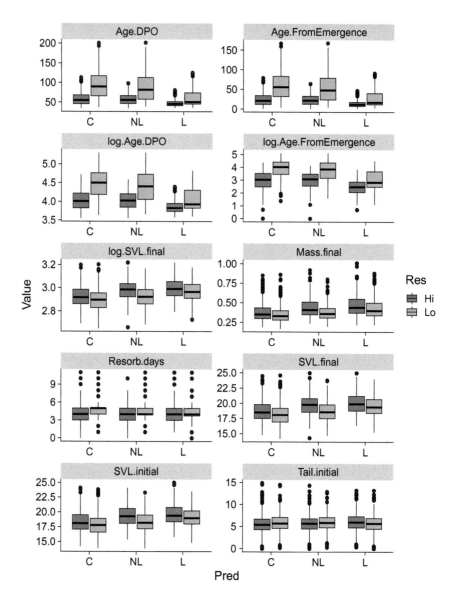

Figure 9.17 With just six lines of code, we made a boxplot for each of the variables in our dataset, plotted for each combination of two different resource levels and our three predator treatments. Not too shabby.

chunk of code, we can easily gather our data into a long format, calculate the means and standard errors for *all of our variables* and then pipe it to ggplot where we will plot a series of bar graphs with error bars. The code here might look a little daunting, but it is nothing more than a collection of things covered throughout this chapter (Figure 9.18).

```
RxP.clean %>%
  gather(key = Measurement, value = Value,
         -Ind, -Block, -Tank, -Tank.Unique,
         -Hatch, -Pred, -Res) %>%
  group_by(Measurement, Pred, Res) %>%
  summarize(Mean = mean(Value),
            SD = sd(Value),
            N = length(Value)) %>%
  mutate(SE = SD/sqrt(N)) %>%
  ggplot(data=., aes(x=Pred, y=Mean, fill=Res))+
  geom_col(position="dodge")+
  facet_wrap(facets=.~Measurement, scales="free", nrow=5)+
  geom_errorbar(aes(ymax = Mean + SE,
                    ymin = Mean - SE,
                    width=0.4),
                position=position_dodge(width=0.9))+
  xlab("Predator treatment")+
  ylab("Trait mean ± SE")+
  scale_fill_manual(values=c("seagreen","sky blue"),
                    name="Resource treatment",
                    labels=c("High","Low"))
```

⊢ Box 9.6 - Take-home message ⊢

- The **tidyverse** has a lot of packages that work together beautifully, like **dplyr** and **ggplot2**.
- **dplyr** is useful for data wrangling and helps you elegantly and efficiently summarize and manipulate your data.
- The functions **gather()** and **spread()** are invaluable for manipulating your data, allowing you to switch between wide and long formats as you need.
- The function **do()** allows you to combine the grouping aspects of **dplyr** with running functions on your data, such as data analysis.
- **qplot()** and **ggplot()** are excellent packages for creating professional figures. These functions give you near total control over your figures and there are practically infinite ways to customize your plots.
- Combining data wrangling and plotting is a powerful technique to quickly go from raw data to summarized figures.

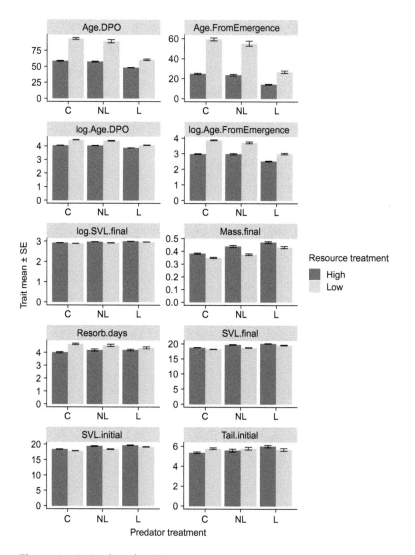

Figure 9.18 Look at that!!!!

9.9 ASSIGNMENT!

Here are some things to do on your own, to build on the data wrangling and plotting skills you have just been working on. Remember you can find example answers and code at https://github.com/jtouchon/Applied-Statistics-with-R.

Your goal is to pick a variable (final SVL, time to resorb the tail, etc.) and make a bargraph that shows the mean and standard error for each tank in the experiment. Make sure to give it meaningful axis titles and all that. Good luck!

Writing Loops and Functions in R

10 Writing Loops and Functions in R

The purpose of this chapter is to introduce you to using R to automate repetitive tasks by writing for loops and custom functions. You do not need to have completed any other chapters in order for this material to be useful, but of course familiarity with R is pretty important at this point, as is familiarity with the RxP dataset.

We will cover the following topics:

- for loops
- writing your own functions

Why are we going to cover this now? Some of this material might seem pretty advanced but I think if you stick with it you will see how it is useful. The main reason that it is helpful to learn about loops and functions is that there are many times when it is a good idea to automate some sort of task that you are going to do over and over.

Applied Statistics with R: A Practical Guide for the Life Sciences. Justin C. Touchon, Oxford University Press (2021). © Justin C. Touchon. DOI: 10.1093/oso/9780198869979.003.0010

10.1 FOR LOOPS

Probably the simplest way to automate completing some task is to use a **for** loop, which is simply a command that specifies a task (or tasks) to be completed a set number of times. Essentially, a for loop runs as follows:

```
for(i in *vector of number of times to do task*)
{
    *task to be completed*
    print(*return some output to the console*)
}
```

For example, imagine you are doing a study with 20 mice and you want to make a list of the mice in your study (by number) you could easily use a for loop to generate it very quickly. The important things to note in the following code are as follows:

1. We have defined a variable called "*i*" and set it equal to the values 1 through 20, so, the action happens 20 times. Whatever we put in between the curly braces (these things: {}) will happen when *i* is equal to 1, then when *i* is equal to 2, then 3, and so on, until 20. Note that "*i*" does not have to be called "*i*," it could be called anything. If you want to code "*for(Susie in 1:20)*," you can, you just have to make sure to substitute the word "*Susie*" in your for loop.
2. We included the function "**print()**" to make sure that at the end of each loop the output was printed to the console.

```
for(i in 1:20)
{
  temp<-paste("mouse",i)
  print(temp)
}
```

```
## [1] "mouse 1"
## [1] "mouse 2"
## [1] "mouse 3"
```

```
## [1] "mouse 4"
## [1] "mouse 5"
## [1] "mouse 6"
## [1] "mouse 7"
## [1] "mouse 8"
## [1] "mouse 9"
## [1] "mouse 10"
## [1] "mouse 11"
## [1] "mouse 12"
## [1] "mouse 13"
## [1] "mouse 14"
## [1] "mouse 15"
## [1] "mouse 16"
## [1] "mouse 17"
## [1] "mouse 18"
## [1] "mouse 19"
## [1] "mouse 20"
```

So, what *exactly* does this code do? First, we set *i* equal to 1. Second, we execute a function called **paste()** which simply pastes two bits of text together and then we store the product of that function in an item called "*temp*." Here, we pasted together the word "*mouse*" and whatever *i* is currently set at. Third, we print what is currently stored in "*temp*" to the console with the "*print(temp)*" command. When the loop gets to the end of executing those three things, it sets *i* equal to 2 and repeats everything, and so on, until the end of our vector of values for *i*, which is 20.

It is often useful to not have to define everything exactly, but instead to simply define the number of loops to run based on some other pre-existing variable. For example, in the RxP study we were studying aspects of meta-morphosis in amphibians. Without even knowing how many observations there were, we can define a for loop that runs for as long as our data frame is. In the next example we have set the for loop to run for as many rows as are in the *RxP.clean* data frame. Another advantage of this method is that if we decide to change our data at all (e.g., remove some outliers), we can rerun this code without modifying a thing.

```
for(i in 1:nrow(RxP.clean))
{
    temp<-paste("I am tadpole",i)
    print(temp)
}
```

Loops are certainly useful to know about but have actually been somewhat obviated by the many of the functions found in the *dplyr* package. For
example, everything we did with the **do()** function in Chapter 9 could have
been done with for loops, but the **do()** function and the *tidyverse* in general
makes it much easier, faster, and more elegant to accomplish.

10.2 UNDERSTANDING FUNCTIONS

It may be obvious, but nearly everything you do in R is done by a function.
Want to read in your data? Use the **read.csv()** function. Want to calculate
a mean? Use the **mean()** function. Want to run a generalized linear mixed
effects model? Better use the **glmer()** or **glmmTMB()** functions.

Functions are sets of commands that are bundled together in order to
accomplish some task. Generally, but not always, functions require some
input(s) and return some sort of output(s). This can be as simple as the
mean() function, which simply requires a vector of numbers, all the way up
to something like the complex statistical models run throughout this book.
Writing functions and loops requires a little bit of an open mind and an
ability to think in the abstract. That said, once you wrap your head around
it, it can really open up a lot of possibilities for you.

Box 10.1 - Why write functions?

Although R may seem like it has a function for every possible task, it does not. Thus,
there are many reasons why you might want to create your own functions. For example,
you might want to write a function for any of the following tasks:

- Editing data
- Repeating similar analyses on different variables
- Creating similar figures from different variables
- Creating customized figures that require a lot of code
- Running simulations
- Any time you are doing something over and over …

10.3 WRITING FUNCTIONS

Functions are similar to loops in that they can be very useful for automating repetitive tasks, but they are also considerably different. Loops execute a specific task many times. They are essentially the same as if you just typed out the commands over and over. Thus, objects that are created within loops are stored in the memory. In the previous examples, the "*temp*" object will be found in your memory and will be the last version of it that the loop created. In functions, objects that are created internally to the function *never exist outside the function*, unless you specify them to. Thus, you can have a very complex function that defines many temporary objects, but those objects will not exist outside the function and therefore will not clutter up your workspace.

The other major difference is that functions are generally designed to have flexible input, so you can have a single function that you can do multiple things with. This all depends of course on how you build your function and what arguments you build into it.

Box 10.2 - General advice for writing functions

When you want to write a function, it is a good idea to sit down with pen and paper and think about what you want the function to do. It is also helpful to think of it as an iterative process. Start with a very simple function and build from there. One major benefit of this is it helps you get any kinks worked out as you expand your function and it grows in complexity. It is also a very good idea to very thoroughly annotate your functions with descriptions of what different lines are doing.

10.4 HOW A FUNCTION WORKS

In the most basic form, a function looks more or less like this:

```
function_name = function(argument1, argument2…)
{
  #annotation goes here
  commands go here
  return(output goes here)
```

Notice that, instead of ending with **print()** like in the loop, we end with **return()**. Whatever you specify in the **return()** function is what the function will output. As mentioned previously, any other variables or objects you define will not exist outside of the function.

Okay, before we actually make a function, I just want to repeat something. Start with the simplest version of what you might want to do, then build up from there! It is easy to dive in and get excited, and then get stuck in some quicksand and not know how to get out. Always start simple and build from there.

10.4.1 A realistic and useful function: Let's calculate the standard error

While there are a million reasons you might want to write a function, one of the most obvious is because you need to do something that the R gurus haven't already written. A classic example is calculating standard error. While it is something that many of us have to do from time to time, it is not a function that is actually built into R. Go figure!

Despite its absence, we can easily write our own function to calculate the standard error. It's a simple bit of math, right? Just the standard deviation of a group of numbers divided by the square root of the count of those numbers, aka, the sample size.

In the next section, we've created a new function called "*stderr*" which takes a single input called "*vec*," which should be a vector of numbers. The function does the simple math and returns the standard error of that vector. Easy peasy.

```
stderr<-function(vec)
{
    return(sd(vec)/sqrt(length(vec)))
}
```

It's important to recognize what is going on in that code. The name "*vec*" which is in the parentheses after "*function*" is a completely arbitrary placeholder name. It just signifies that whatever *you place in the parentheses when you execute the function* will then be called "*vec*" internally in the function and will be used everywhere you coded "*vec*" to be used. So, when we execute the function the vector of numbers in the parentheses will be placed into where we see "*vec*" in the function. The standard deviation of the vector will be calculated via the **sd()** function, which will be divided by the square root of the length of the vector. Whatever is calculated will then be returned to the command prompt via the **return()** function.

We can test that it works by providing the function with a vector of numbers, such as 1 through 50.

```
stderr(1:50)
```

```
## [1] 2.061553
```

We can even incorporate our new function into older code we had where we were summarizing our data. For example, in an earlier chapter we calculated the standard error of the final SVL of our metamorphs so that we could use it in plotting. Now, we can just use our handy dandy function to do that job.

```
RxP.byTank %>%
      group_by(Res, Pred) %>%
      summarize(SVL.mean = mean(SVL.final),
                SVL.se = stderr(SVL.final))
```

```
## # A tibble: 6 x 4
## # Groups:    Res [2]
##    Res    Pred   SVL.mean  SVL.se
##    <chr>  <fct>     <dbl>   <dbl>
## 1 Hi     C          18.8   0.162
## 2 Hi     NL         19.4   0.420
## 3 Hi     L          20.5   0.303
## 4 Lo     C          18.2   0.184
## 5 Lo     NL         18.7   0.315
## 6 Lo     L          20.4   0.415
```

10.5 WRITING MORE COMPLEX FUNCTIONS: AN EXAMPLE USING SIMULATIONS

As I've already said, there are many times when you might want to write your own functions. As a scientist, one of the most useful things you can do is simulate the data you might get as a way to understand the effects of sample size and effect size on your statistical power. For example, you can use a pilot study to estimate what you think the effect size of your treatments might be and then simulate data to figure out how big a sample size you will need to observe statistical significance. Maybe this means you have pilot data on two drugs you are testing out, or two behavioral treatments you are administering, or two temperatures you are raising plants or worms or whatever in. As long as you have *some idea* of what the effect size might be (i.e., the difference in the means of your treatments), the examples we will go through here will be useful to you.

Imagine we are conducting predation trials with two different predators and we want to know how many replicates we need to run in order to know if there is a significant difference between them. Let's write a function to simulate the effects of the two predators. It's important to recognize that everything we do here can be done for *any type* of variable.

10.5.1 Starting simple

First, we will start very simple. The **rbinom()** function allows you to calculate a binomial probability (i.e., a coin toss) for a given sample size and probability. **rbinom()** takes three arguments.

1. *n*: the number of discrete events you are simulating.
2. *size*: how many times you are running the event at a time.
3. *prob*: the probability of a successful outcome (e.g., a heads in the coin toss).

For example, this code simulates flipping 1 fair coin 10 different times. The output corresponds to each time you flipped a coin, whether it was heads (1) or tails (0).

```
#Flip 1 coin 10 times
rbinom(n=10, size=1, prob=0.5)
```

```
## [1] 0 0 1 1 0 1 0 1 0 0
```

In contrast, this code simulates flipping 10 fair coins 1 time. The output corresponds to how many times you got a heads out of the 10 times you flipped the coin. Notice the difference in these two lines of code. The first one defines the n as 10 and the size as 1, whereas the second one defines the n as 1 and the size as 10.

```
#Flip 10 coins 1 time
rbinom(n=1, size=10, prob=0.5)
```

```
## [1] 3
```

We can also think about this as a predation trial with 10 animals and a predator that eats 50% of prey on average. If you run the last line of code over and over, you will see that you get different answers each time (more or less). Sometimes you get 5, sometimes 4 or 6, less frequently 3 or 7, and so on.

We can easily simulate *lots* of predation trials and then examine them with a histogram (Figure 10.1).

```
#This code simulates 1000 predation trials with a predator
#that kills 50% of prey on average and where each
#predation trial has 10 prey animals.
hist(rbinom(1000,10,0.5))
```

We can also use the function **table()** to calculate the number of each possible outcome we get.

```
table(rbinom(1000,10,0.5))
```

```
##
##   0   1   2   3   4   5   6   7   8   9  10
##   2  11  39 110 221 260 182 123  43   7   2
```

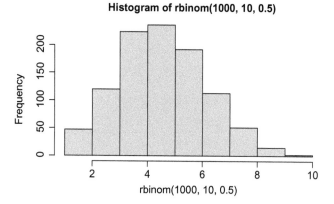

Figure 10.1 A histogram of a 1000 predation trials
simulating a predator that eats, on average, 50% of the prey.
Note that while five prey being consumed is the most
common outcome, it is also very likely that four or six get
eaten, and slightly less likely that three or seven get eaten.

If we divide that output by the number of trials (i.e., 1000) then we can
easily get the proportion of times each result occurred.

```
table(rbinom(1000,10,0.5))/1000
```

```
##
##     1     2     3     4     5     6     7     8     9    10
## 0.008 0.038 0.105 0.177 0.270 0.236 0.122 0.034 0.009 0.001
```

Note, of course, that your outputs to these lines of code will be slightly
different than what is printed here, since my numbers were generated
randomly and so were yours.

10.5.2 Making a very basic function

Let's write a function to calculate the spread of probabilities for two different predators at once. The reason to write this into a function is just to
roll everything up into a single simple package. Plus, it paves the way for
building bigger and better functions!

Notice that the following function has no arguments yet. This is just a function that will execute everything in the curly braces (these things: {}), but we are not passing it any arguments yet. Don't worry, we'll get there. What does this actually do? This code generates 5000 binomial trials with 10 prey in each, with a probability of success (i.e., predation) of 0.3 or 0.5. It stores each of those sets of simulated values in *temporary* objects called "*predA*" and "*predB*." Next, it creates a vector of 10000 letters, the first half of which are "*A*" and the second half of which are "*B*." Perhaps the most important thing to point out is that we use the function **return()** to specify what we want the function to output. Here, we've just specified to make a data frame with two columns, the first is the names of the two predators (*A* or *B*) and the second being the contents of "*predA*" and "*predB*." Lots of things happen inside a function that are only temporary, and they don't exist outside of the function. **return()** is important because it determines what the product of the function will be.

```
comparePred<-function()
{
    predA<-rbinom(5000,10,0.3)
    predB<-rbinom(5000,10,0.5)
    preds<-rep(c("A","B"),each=5000)
    return(data.frame("pred"=preds, "eaten"=c(predA,predB)))
}
```

Now, if you typed that in and nothing happened, you might be wondering why nothing happened. This code merely defined a new function, but it didn't execute it. As with any function, we can execute it directly at the command prompt or assign it to an object. Let's do the latter, shall we?

```
temp<-comparePred()
str(temp)
```

```
## 'data.frame':    10000 obs. of  2 variables:
##  $ pred : chr  "A" "A" "A" "A" ...
##  $ eaten: int  2 1 4 3 2 4 4 6 4 5 ...
```

Okay! That tells us we have a data frame with two columns. The first column tells us which predator we are looking at and the second tells us how many prey they ate. We can visualize the distributions of these simulated trials with **qplot()** (Figure 10.2).

```
qplot(data=temp,
      x=eaten,
      fill=pred,
      geom="histogram",
      facets=pred~.,
      bins=11)
```

How cool! Let's rework and expand this to make it more functional. The following code does several things.

1. First, we build an empty data frame to store our values, making it twice as long as we want the number of simulations.

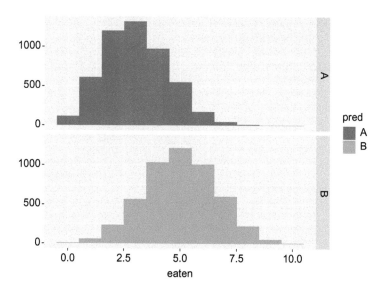

Figure 10.2 Now we have two histograms showing two predators that differ in how many prey they capture, on average. Predator A captures 30% of its prey whereas predator B captures 50%. Once again, since those are the average predation rates there is some degree of likelihood that more or fewer prey get captured in different simulated predation trials.

2. We make a column called "*Pred*," the first half of which will be "*PredA*" and the second half of which will be "*PredB*."

3. Then, we put the 1st predator's data in the 1st half of the data frame, and the 2nd predator's in the 2nd half.

4. We end by returning the whole data frame we created.

```
comparePred<-function()
{
#Build the empty data frame
simData<-data.frame(Trial=rep(1:10,times=2),
                    Pred=rep(c("PredA","PredB"),
                             each=10),
                    Eaten=NA)
#Simulate the predation trials
simData$Eaten[1:10]<-rbinom(10,10,0.3)
simData$Eaten[11:20]<-rbinom(10,10,0.5)
return(simData)
}
```

If we run the function, we can see that it has created a data frame with three columns, one called "*Trial*" which contains the number of the replicate simulated, one called "*Eaten*" which contains the number of prey animals that were eaten during the trial, and one called "*Pred*" which stores the predator in question, A or B.

```
temp<-comparePred()
head(temp)
```

```
##    Trial  Pred Eaten
## 1      1 PredA     4
## 2      2 PredA     5
## 3      3 PredA     4
## 4      4 PredA     2
## 5      5 PredA     2
## 6      6 PredA     1
```

Pretty cool right? At this point it is useful to think about what we can do to make our function more useful. What we have written so far just does one specific task, but the beauty of functions is that they are *flexible*. We can write them so that they take various types of inputs, which is part of why they are more useful than for loops.

How can we make our function more useful, more flexible, more general? I can think of three obvious things we might want to do with a function like this. We might want to:

1. Add in the ability to change the number of trials.
2. Add in the ability to change the level of predation.
3. Add in a statistical test to evaluate the difference between the two predators.

Let's do each of these in succession, making our function more useful (and complex) as we go.

10.5.3 Building in arguments

First, let's start by adding a single argument to specify the number of trials we want to simulate. In our original function, we specified that we wanted to run 10 trials. But why should that be fixed? Perhaps we want to explore how our ability to detect a difference in our predators changes with the number of trials we run? I will point out here that this is essentially what a power analysis does, but we are building it from the ground up.

In the following code, we've specified a single argument which will just be a number—the number of trials we want to simulate running. Then in the code, we've used the name of that argument ("*numTrials*") as the placeholder for everywhere we would have previously specified running 10 trials. We make the "*Trial*" column of the data frame go from 1 until whatever number "*numTrials*" is and we run each **rbinom()** function for as many times as "*numTrials*" is. However, since the size of our data frame will change with the size of whatever "*numTrials*" is specified to be, we can't hard code the rows to fill like we did before. Thus, we use square brackets to subset the data frame to when Pred is equal to "*PredA*" or "*PredB*."

```
comparePred<-function(numTrials)
{
    #Build the empty data frame
    simData<-data.frame(Trial=rep(1:numTrials,times=2),
                        Pred=rep(c("PredA","PredB"),
                                 each=numTrials),
                        Eaten=NA)
  #Simulate the predation trials
    PredA_trials<-rbinom(numTrials,10,0.3)
    PredB_trials<-rbinom(numTrials,10,0.5)
    simData$Eaten[simData$Pred=="PredA"]<-PredA_trials
    simData$Eaten[simData$Pred=="PredB"]<-PredB_trials
    return(simData)
}
```

Okay, we've built in the ability to vary the number of trials and the code works. If you run it and vary the number of trials, you will get some results that you may or may not care about. Now, let's add in the ability to change the level of predation. Currently, we've hard coded that our predators eat 30% and 50% of prey, on average. But what if we want to explore how the difference between our predators affects things?

In the following code, we've added in two new arguments, titled "*predLevelA*" and "*predLevelB*." Each of these will be a probability of the predator successfully getting its prey, between 0 and 1. Everything in the code is the same as the last iteration, except now the predation levels are determined by whatever we want them to be. In each of the **rbinom()** functions, we've replaced 0.3 and 0.5 with the values we will set when we run the function.

```
comparePred<-function(numTrials,predLevelA, predLevelB)
{
    #Build the empty data frame
    simData<-data.frame(Trial=rep(1:numTrials,times=2),
                        Pred=rep(c("PredA","PredB"),
                                 each=numTrials),
                        Eaten=NA)
```

```
#Simulate the predation trials
  PredA_trials<-rbinom(numTrials,10,predLevelA)
  PredB_trials<-rbinom(numTrials,10,predLevelB)
  simData$Eaten[simData$Pred=="PredA"]<-PredA_trials
  simData$Eaten[simData$Pred=="PredB"]<-PredB_trials
  return(simData)
}
```

Let's see our function in action! We can run our function twice and see how changing the difference in predation between our two predators effects our general ability to see a difference between them.

```
temp.1<-comparePred(20, 0.2, 0.3)
temp.4<-comparePred(20, 0.2, 0.6)
plot.1<-qplot(data=temp.1,
              x=Eaten,
              fill=Pred,
              geom="histogram",
              facets=Pred~.,
              main="10% difference",
              bins=10)
plot.4<-qplot(data=temp.4,
              x=Eaten,
              fill=Pred,
              geom="histogram",
              facets=Pred~.,
              main="40% difference",
              bins=10)
plot_grid(plot.1, plot.4, ncol=2)
```

We can see pretty easily that a larger difference in predation level makes it easier to visually detect a difference in our predators (Figure 10.3). But what about a statistical difference? Without much difficulty, we can modify our function to output not the actual data frame, but instead the results of a statistical comparison between the two predators.

Imagine you are trying to design a new experiment. Maybe you are writing a grant proposal. Maybe you've already got the grant proposal funded! Either way, you are planning out an experiment. Hopefully you have some idea of what *might* happen. A simulation like the following will allow you to estimate with confidence how big of a study you should do if you want to obtain significant results!

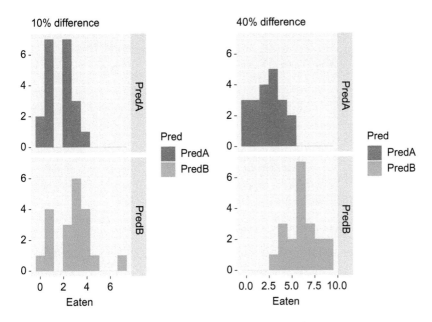

Figure 10.3 Wow, now we are really cooking! These figures demonstrate how we can much more easily discriminate two hypothetical predators when the difference between them is greater. That may be obvious, but it is important to keep in mind because it shows how small differences, even if real, can be hard to detect with experiments and statistics.

This code is just a modification of what we've been doing. The arguments are all the same, but we've added a new blank column at the start called "*NotEaten*" and we've filled it in with the number of survivors from each simulated predation trial. At the end, we've added a new section where we run a binomial generalized linear model testing for a statistical difference between our two simulated predators. Lastly, we return not the data frame, but just the p-value from the GLM, as calculated by **Anova()** from the *car* package.

```
predTrialSim<-function(numTrials,predLevelA,predLevelB)
{
    #Build the empty data frame
    simData<-data.frame(Trial=rep(1:numTrials,times=2),
                        Pred=rep(c("PredA","PredB"),
                             each=numTrials),
```

```
                            Eaten=NA,
                            NotEaten=NA)
    #Simulate the predation trials
    PredA_trials<-rbinom(numTrials,10,predLevelA)
    PredB_trials<-rbinom(numTrials,10,predLevelB)
    simData$Eaten[simData$Pred=="PredA"]<-PredA_trials
    simData$Eaten[simData$Pred=="PredB"]<-PredB_trials
    simData$NotEaten<-10-simData$Eaten
    #run the GLM and return just the p-value
    simGLM<-glm(cbind(Eaten,NotEaten)~Pred,
                data=simData,
                family="binomial")
    return(Anova(simGLM)$Pr)
}
```

If we run this function a bunch of times, we see that we get just a *P*-value out, and that it varies from run to run. Sometimes it might be significant (smaller than 0.05) and sometimes not. Note of course, that whatever values you obtain will be different than what is printed here, since all of these numbers are generated randomly.

```
predTrialSim(20,0.2,0.3)
```

```
## [1] 0.03597159
```

```
predTrialSim(20,0.2,0.3)
```

```
## [1] 0.01862498
```

```
predTrialSim(20,0.2,0.3)
```

```
## [1] 0.005096244
```

```
predTrialSim(20,0.2,0.3)
```

```
## [1] 0.01808329
```

```
predTrialSim(20,0.2,0.3)
```

```
## [1] 1.920063e-05
```

```
predTrialSim(20,0.2,0.3)
```

```
## [1] 0.8246796
```

At this point, we have made a function that simulates running predation trials by two predators and allows us to vary how many trials we run and how effective the predators are, and we conduct a statistical test to see if our hypothetical experiments have yielded significant results or not. In order to glean anything truly meaningful from such a simulation, we need to run it *lots* of times. That is where the **replicate()** function comes in. **replicate()** allows you to run a function however many times you want, and it returns the output as a vector. Now we can use **replicate()** to run our function many, many times and see what the general pattern is as far as our ability to detect statistically significant differences between predators. Let's imagine a fairly small difference between our predators—say 10%—and let's explore how running 5, 10, 20, and 40 trials impacts our ability to detect a difference between them.

```
predSims.5<-replicate(1000,predTrialSim(5,0.5,0.6))
predSims.10<-replicate(1000,predTrialSim(10,0.5,0.6))
predSims.20<-replicate(1000,predTrialSim(20,0.5,0.6))
predSims.40<-replicate(1000,predTrialSim(40,0.5,0.6))
```

Now let's plot those results. If we combine the output from each simulation into a single data frame, we can plot it using **qplot()** and facet by the number of trials run. The following first line combines each of the four simulations into a single data frame and creates a column called "*Reps*" that is the number of replicates used in each simulation. The function **geom_vline()** adds a red vertical line at 0.05 on the x-axis, which will allow

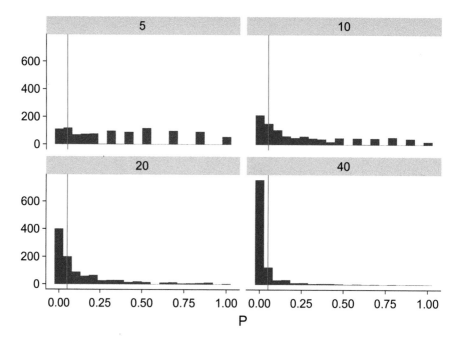

Figure 10.4 Here we can clearly see how running more replicates of our experiment increases the probability of detecting a difference. Recall that, in each of these panels, we were trying to detect a statistically significant difference between two predators that differ in their predation rate by 10%. In the case where we only run five replicates, we are very unlikely to detect a difference, whereas running 40 replicates means we are almost certain to detect the 10% difference between the predators.

us to easily see the cutoff (however arbitrary) between significant and non-significant differences (Figure 10.4).

```
#Combine simulations into a single data frame
predSims<-data.frame(Reps=rep(c(5,10,20,40),
                          each=1000),
                     P=c(predSims.5,
                         predSims.10,
                         predSims.20,
                         predSims.40))
#Plot
qplot(data=predSims, x=P, geom="histogram", bins=20)+
  facet_wrap(facets=Reps~.)+
  geom_vline(xintercept=0.05, color="red")+
  theme_cowplot()
```

As we increase our number of trials run, we can clearly see that our ability to detect a statistically significant difference between our predators increases. Recall that we *know* there is a difference, it's 10% on average. But, that is different than a *statistical* difference. If we only run 5 predation trials, we are very unlikely to mathematically detect a difference. We can easily calculate what percentage of simulated *P*-values are below 0.05. The following code works because of two things.

1. We are making logical statements asking if the values in each of the different "*predSims*" vectors are smaller than 0.05.
2. Logical statements return True or False, and True is equal to 1 whereas False is equal to 0, allowing us to easily sum the number of True responses and divide it by the total number of trials we ran.

```
sum(predSims.5<0.05)/1000
```

```
## [1] 0.183
```

```
sum(predSims.10<0.05)/1000
```

```
## [1] 0.317
```

```
sum(predSims.20<0.05)/1000
```

```
## [1] 0.513
```

```
sum(predSims.40<0.05)/1000
```

```
## [1] 0.802
```

If you were designing this study, how many replicates would you run? If you were to run 20, there would be about a 50% chance you would detect statistical significance. But, if you run 40, you have about an 80% chance of detecting a difference between the predators.

Box 10.3 - Take home message

Hopefully this has helped to somewhat demystify writing loops and functions and demonstrate that it really is not that hard. If we wanted, we could use our *"predTrialSim"* function to build another function that takes a vector of differences between predators or a vector of sample sizes and would run all the models and spit it out in a single table for us to interpret. Wouldn't that be something? The other thing I hope I've demonstrated is the power of simulating your data. If you have some inkling of how different your treatments might be, you can simulate data to get a good idea of how many trials you might need to run. That is extremely powerful.
Remember:

- Start small and build up from there. If you just try to dive right into a big complex function you are more likely to make mistakes and get very frustrated.
- Taking the time to plan and think carefully about what you want your function to do can be invaluable. Think about what you need the function to do and walk through how you will do it. Do you need to make a data frame or matrix to store values? Do you need to use a for loop to cycle through something?
- Don't forget to include the **return()** function at the end.

10.6 ASSIGNMENT!

Here is an assignment to work on the skills you have just developed. Answers to these problems and the associated code can be found on Github https://github.com/jtouchon/Applied-Statistics-with-R.

1. Modify the final version of the function we wrote (*predTrialSim*) to include another argument which allows you to vary the number of prey animals in each predation trial.
2. Pick at least four values for the number of prey in each trial and make a meaningful plot which explores how detecting a statistically significant difference in predators varies with the size of the pool of animals in each trial.
3. Which is more important, running more trials or running trials with more animals?

Final Thoughts

11 Final Thoughts

Well, here we are. I hope that you have learned a lot on your journey through this book and feel comfortable and confident analyzing your data. I want to take a few moments to recap what we have covered during the last ten chapters.

11.1 UNDERSTANDING YOUR DATA IS THE MOST IMPORTANT PRECURSOR TO ANALYZING IT

The single most important thing you need to do when preparing for your analyses is to understand the structure of your data. This simple piece of advice will make a world of difference. Are your data normally distributed or not? If they are not, what is the best error distribution to approximate them? Are there random affects you need to control for? These are the sorts of things you need to invest time in understanding before you ever code an analysis.

Applied Statistics with R: A Practical Guide for the Life Sciences. Justin C. Touchon, Oxford University Press (2021). © Justin C. Touchon. DOI: 10.1093/oso/9780198869979.003.0011

11.2 KNOWING HOW TO GET HELP IS ESSENTIAL

If you've invested in this book you are clearly interested in learning more about R and how to analyze your data. But *no book can teach you everything you will need to know*. R is a living breathing language that is constantly changing, evolving, growing, and being updated. Just like learning how to read R help files is a skill that takes time to develop, so does finding help online. The internet is your friend, but there is a lot of information out there and it is not always written in a way that is particularly user-friendly.

Hopefully while working through this book you have been able to apply the code I've written to your own data. One of the most important things you can learn how to do is to take some bit of code that is posted online and written for someone else's data and modify it for your own use. If you get an error message when you are trying to run a model, whether it is a generalized linear model (GLM) or a mixed-effects model, your first step is to copy the error and search for it on the internet. This may not be helpful, but hopefully it will be. Many relatively simple errors have fairly inscrutable messages that get returned to the console. You will have to learn what these mean, which takes time. But take heart that the internet is a great resource for getting help.

11.3 YOUR DATA ANALYSIS SHOULD BE CLEAR FROM THE OUTSET AND YOU SHOULD AVOID QUESTIONABLE TECHNIQUES

When you embark on analyzing your data, make sure you have a deliberate and clear goal in mind. As mentioned already, you should understand the structure of your data, and that should guide the sorts of analyses you are doing. But just as important, you should have a clear idea of what the predictor variables are and what questions you are asking with your analyses. Do not embark on fishing expeditions where you throw every imaginable analysis or predictor at a set of response variables in the hope that you find something "significant" which will make your advisor or your manuscript reviewers happy.

Remember that statistical significance is an artificial construct, a line in the sand that we've come to accept but which is not necessarily more or less meaningful than other possible lines in the sand. The important thing is understanding if your results are important and meaningful. The more data you have, the easier it is to detect "significant" effects that are essentially meaningless. Always make sure that what you are reporting makes sense.

11.4 PRESENTING YOUR DATA IN WELL-CONSTRUCTED FIGURES IS KEY

R is an amazing program for making figures and I hope I've been able to give you some of the tools you will need to do just that. Making effective figures is one of the most powerful tools you can have at your disposal. Figures should accurately represent both the nature of your data but also visually convey the analyses you have conducted. Good figures help the reader understand what it is that you have spent months agonizing over in just a few moments. Always remember to make your axes large enough for folks to easily read and use color well (not too much but not too little).

Make sure to check out the GitHub page for this book (https://github.com/jtouchon/Applied-Statistics-with-R) in order to access updates and find answers to the exercises at the end of each chapter.

Happy coding! I wish you the best of luck with your data analysis.

Index

$, 54
<-, 10
[], 16, 155
%>%, 70, 237

A

abline(), 158
aggregate(), 69
AIC(), 115, 187
analysis of covariance, 163
analysis of variance, 125, 141
ANCOVA, 163, 200
annotation, 4
ANOVA, 125, 141
Anova(), 124, 142, 195
anova(), 124, 217
arguments, 13
assign operator, 10

B

balance, 37
bar graphs, 89
barplot2(), 89
best practices, 42
blocked designs, 39
boxplot, 56, 81
broom package, 237

C

c(), 17
cld(), 132, 146, 151
colors, 78
concatenate function, 17
confidence intervals, 168
cowplot package, 77

creating objects, 9
crossed random effects, 211

D

data dredging, 42
data frame, 19
data wrangling, 235
dependent variable, 45
diagnostic plots, 129, 189, 215
do(), 249
dplyr package, 62, 74, 235

E

emmeans package, 231
emmeans(), 132, 146, 151, 165
emtrends(), 165
error distribution, 114
error distributions, 181
expand.grid(), 172, 204
experimental design, 37

F

filter(), 242
fitdistr(), 115, 187
fixed effects, 209
for loops, 285
full_join(), 243
functions, 13

G

gather(), 245
generalized linear models, 181
geom_smooth(), 161
ggplot(), 174, 274
ggplot2 package, 77, 80, 96, 161, 235, 256

ggplot2 themes, 270
glht(), 132
glimpse(), 185
glm(), 181
glmer(), 210
glmmadmb package, 210
glmmTMB package, 210
glmmTMB(), 221
group_by(), 74, 237, 238

H

HARKing, 42
help function, 13

I

independent variable, 45
indexing, 16, 29, 60, 155
interactions, 143

J

joining data, 243

K

Kruskal-Wallis test, 118

L

labeller(), 273
left_join(), 243
likelihood ratio test, 217
linear model, 105, 124, 139
linear regression, 153
link functions, 181
literate programming, 4
lm(), 105, 124, 139, 153
lme4 package, 210
lmer(), 210
logical statements, 61
logistic regression, 198

M

mann-whitney U test, 118
MASS package, 187
MCMCglmm package, 210
mixed effects models, 209
model coefficients, 126, 144
model reduction, 201
model selection, 201, 227
multiple comparisons, 132, 146, 165

mutate(), 74, 240

N

nested models, 217
nested random effects, 211
non-normal data, 114, 181
non-parametric statistics, 117
normal data, 104, 114, 139
normality tests, 109

O

object names, 11
organizing your data, 21
overdispersion, 192

P

p-hacking, 42
packages, 2
parametric statistics, 104
pipe, 70
pipe function, 237
plot(), 124
post-hoc tests, 132, 146, 151, 165
predict(), 168, 204
predictor variable, 45
pseudoreplication, 107

Q

Q-Q plot, 130, 189, 215
qplot(), 80, 161, 258, 274

R

R as a calculator, 12
random effects, 209
random intercept, 211
random slope, 211
randomization, 37
read.csv(), 51
reordering factor levels, 72
replication, 37
reproducible code, 8
residuals, 129
response variable, 45
restricted maximum likelihood, 227
RGB colors, 79
right_join(), 243
RStudio, 2, 52
RxP, 49, 106

S

scatterplots, 87
script window, 3
select(), 241
shapiro.test(), 112
simulations, 292
slice(), 157
spread(), 245
square brackets, 16, 155
Student's t-test, 120
subset, 60
summarize(), 69, 70, 74, 238
summary(), 29, 124,
    142, 214

T

t.test(), 120
tidyr package, 235
tidyverse, 62, 235
Tukey's HSD, 132, 146, 165

V

vectors, 15
violin plots, 84

W

write.csv(), 109
writing functions, 285, 288, 298